Teubner Skripten zur
Mathematischen Stochastik

Klaus D. Schmidt
Lectures on Risk Theory

Teubner Skripten zur Mathematischen Stochastik

Herausgegeben von
Prof. Dr. rer. nat. Jürgen Lehn, Technische Hochschule Darmstadt
Prof. Dr. rer. nat. Norbert Schmitz, Universität Münster
Prof. Dr. phil. nat. Wolfgang Weil, Universität Karlsruhe

Die Texte dieser Reihe wenden sich an fortgeschrittene Studenten, junge Wissenschaftler und Dozenten der Mathematischen Stochastik. Sie dienen einerseits der Orientierung über neue Teilgebiete und ermöglichen die rasche Einarbeitung in neuartige Methoden und Denkweisen; insbesondere werden Überblicke über Gebiete gegeben, für die umfassende Lehrbücher noch ausstehen. Andererseits werden auch klassische Themen unter speziellen Gesichtspunkten behandelt. Ihr Charakter als Skripten, die nicht auf Vollständigkeit bedacht sein müssen, erlaubt es, bei der Stoffauswahl und Darstellung die Lebendigkeit und Originalität von Vorlesungen und Seminaren beizubehalten und so weitergehende Studien anzuregen und zu erleichtern.

Lectures on Risk Theory

By Prof. Dr. sc. math. Klaus D. Schmidt
Techn. Universität Dresden

 B. G. Teubner Stuttgart 1996

Prof. Dr. sc. math. Klaus D. Schmidt

Born 1951 in Glückstadt, studies of mathematics 1970–1975 at Universität Kiel and Universität Zürich, PhD 1980 and habilitation 1988 at Universität Mannheim, since 1993 professor of mathematics at Technische Universität Dresden.

Die Deutsche Bibliothek – CIP-Einheitsaufnahme

Schmidt, Klaus D.:
Lectures on risk theory / Klaus D. Schmidt. – Stuttgart:
Teubner, 1996
(Teubner Skripten zur mathematischen Stochastik)

ISBN 978-3-519-02735-5   ISBN 978-3-322-90570-3 (eBook)
DOI 10.1007/978-3-322-90570-3

Herstellung: Druckhaus Beltz, Hemsbach/Bergstraße

To the memory of

Peter Hess

and

Horand Störmer

Preface

Twenty–five years ago, Hans Bühlmann published his famous monograph *Mathematical Methods in Risk Theory* in the series *Grundlehren der Mathematischen Wissenschaften* and thus established nonlife actuarial mathematics as a recognized subject of probability theory and statistics with a glance towards economics. This book was my guide to the subject when I gave my first course on nonlife actuarial mathematics in Summer 1988, but at the same time I tried to incorporate into my lectures parts of the rapidly growing literature in this area which to a large extent was inspired by Bühlmann's book.

The present book is entirely devoted to a single topic of risk theory: Its subject is the development in time of a fixed portfolio of risks. The book thus concentrates on the claim number process and its relatives, the claim arrival process, the aggregate claims process, the risk process, and the reserve process. Particular emphasis is laid on characterizations of various classes of claim number processes, which provide alternative criteria for model selection, and on their relation to the trinity of the binomial, Poisson, and negativebinomial distributions. Special attention is also paid to the mixed Poisson process, which is a useful model in many applications, to the problems of thinning, decomposition, and superposition of risk processes, which are important with regard to reinsurance, and to the role of martingales, which occur in a natural way in canonical situations. Of course, there is no risk theory without ruin theory, but ruin theory is only a marginal subject in this book.

The book is based on lectures held at Technische Hochschule Darmstadt and later at Technische Universität Dresden. In order to raise interest in actuarial mathematics at an early stage, these lectures were designed for students having a solid background in measure and integration theory and in probability theory, but advanced topics like stochastic processes were not required as a prerequisite. As a result, the book starts from first principles and develops the basic theory of risk processes in a systematic manner and with proofs given in great detail. It is hoped that the reader reaching the end will have acquired some insight and technical competence which are useful also in other topics of risk theory and, more generally, in other areas of applied probability theory.

I am deeply indebted to Jürgen Lehn for provoking my interest in actuarial mathematics at a time when vector measures rather than probability measures were on my mind. During the preparation of the book, I benefitted a lot from critical remarks and suggestions from students, colleagues, and friends, and I would like to express my gratitude to Peter Amrhein, Lutz Küsters, and Gerd Waldschaks (Universität Mannheim) and to Tobias Franke, Klaus–Thomas Heß, Wolfgang Macht, Beatrice Mensch, Lothar Partzsch, and Anja Voss (Technische Universität Dresden) for the various discussions we had. I am equally grateful to Norbert Schmitz for several comments which helped to improve the exposition.

Last, but not least, I would like to thank the editors and the publishers for accepting these *Lectures on Risk Theory* in the series *Skripten zur Mathematischen Stochastik* and for their patience, knowing that an author's estimate of the time needed to complete his work has to be doubled in order to be realistic.

Dresden, December 18, 1995   Klaus D. Schmidt

Contents

Introduction   1

1   **The Claim Arrival Process**   **5**
   1.1   The Model   5
   1.2   The Erlang Case   9
   1.3   A Characterization of the Exponential Distribution   12
   1.4   Remarks   16

2   **The Claim Number Process**   **17**
   2.1   The Model   17
   2.2   The Erlang Case   21
   2.3   A Characterization of the Poisson Process   23
   2.4   Remarks   41

3   **The Claim Number Process as a Markov Process**   **43**
   3.1   The Model   43
   3.2   A Characterization of Regularity   51
   3.3   A Characterization of the Inhomogeneous Poisson Process   56
   3.4   A Characterization of Homogeneity   62
   3.5   A Characterization of the Poisson Process   76
   3.6   A Claim Number Process with Contagion   77
   3.7   Remarks   84

4   **The Mixed Claim Number Process**   **85**
   4.1   The Model   85
   4.2   The Mixed Poisson Process   87
   4.3   The Pólya–Lundberg Process   93
   4.4   Remarks 100

5   **The Aggregate Claims Process**   **103**
   5.1   The Model 103
   5.2   Compound Distributions 109
   5.3   A Characterization of the Binomial, Poisson, and Negativebinomial
        Distributions 115
   5.4   The Recursions of Panjer and DePril 119
   5.5   Remarks 124

6   The Risk Process in Reinsurance   **127**
   6.1   The Model 127
   6.2   Thinning a Risk Process 128
   6.3   Decomposition of a Poisson Risk Process 133
   6.4   Superposition of Poisson Risk Processes 141
   6.5   Remarks 154

7   The Reserve Process and the Ruin Problem   **155**
   7.1   The Model 155
   7.2   Kolmogorov's Inequality for Positive Supermartingales 161
   7.3   Lundberg's Inequality 164
   7.4   On the Existence of a Superadjustment Coefficient 166
   7.5   Remarks 169

Appendix: Special Distributions   **171**
   Auxiliary Notions 171
   Measures 172
   Generalities on Distributions 172
   Discrete Distributions 175
   Continuous Distributions 179

Bibliography   **181**

List of Symbols   **193**

Author Index   **195**

Subject Index   **198**

Introduction

Modelling the development in time of an insurer's portfolio of risks is not an easy task since such models naturally involve various stochastic processes; this is especially true in nonlife insurance where, in constrast with whole life insurance, not only the claim arrival times are random but the claim severities are random as well.

The sequence of claim arrival times and the sequence of claim severities, the claim arrival process and the claim size process, constitute the two components of the risk process describing the development in time of the expenses for the portfolio under consideration. The claim arrival process determines, and is determined by, the claim number process describing the number of claims occurring in any time interval. Since claim numbers are integervalued random variables whereas, in the continuous time model, claim arrival times are realvalued, the claim number process is, in principle, more accessible to statistical considerations.

As a consequence of the equivalence of the claim arrival process and the claim number process, the risk process is determined by the claim number process and the claim size process. The collective point of view in risk theory considers only the arrival time and the severity of a claim produced by the portfolio but neglects the individual risk (or policy) causing the claim. It is therefore not too harmful to assume that the claim severities in the portfolio are i. i. d. so that their distribution can easily be estimated from observations. As noticed by Kupper[1] [1962], this means that the claim number process is much more interesting than the claim size process. Also, Helten and Sterk[2] [1976] pointed out that the separate analysis of the claim number process and the claim size process leads to better estimates of the

---

[1]Kupper [1962]: *Die Schadenversicherung ... basiert auf zwei stochastischen Grössen, der Schadenzahl und der Schadenhöhe. Hier tritt bereits ein fundamentaler Unterschied zur Lebensversicherung zutage, wo die letztere in den weitaus meisten Fällen eine zum voraus festgelegte, feste Zahl darstellt. Die interessantere der beiden Variablen ist die Schadenzahl.*

[2]Helten and Sterk [1976]: *Die Risikotheorie befaßt sich also zunächst nicht direkt mit der stochastischen Gesetzmäßigkeit des Schadenbedarfs, der aus der stochastischen Gesetzmäßigkeit der Schadenhäufigkeit und der Schadenausbreitung resultiert, denn ein Schadenbedarf ... kann ja in sehr verschiedener Weise aus Schadenhäufigkeit und Schadenhöhe resultieren ... Für die K–Haftpflicht zeigt eine Untersuchung von Tröblinger [1975] sehr deutlich, daß eine Aufspaltung des Schadenbedarfs in Schadenhäufigkeit und Schadenhöhe wesentlich zur besseren Schätzung des Schadenbedarfs beitragen kann.*

aggregate claims amount, that is, the (random) sum of all claim severities occurring in some time interval.

The present book is devoted to the claim number process and also, to some extent, to its relatives, the aggregate claims process, the risk process, and the reserve process. The discussion of various classes of claim number processes will be rather detailed since familiarity with a variety of properties of potential models is essential for model selection. Of course, no mathematical model will ever completely match reality, but analyzing models and confronting their properties with observations is an approved way to check assumptions and to acquire more insight into real situations.

The book is organized as follows: We start with the claim arrival process (Chapter 1)

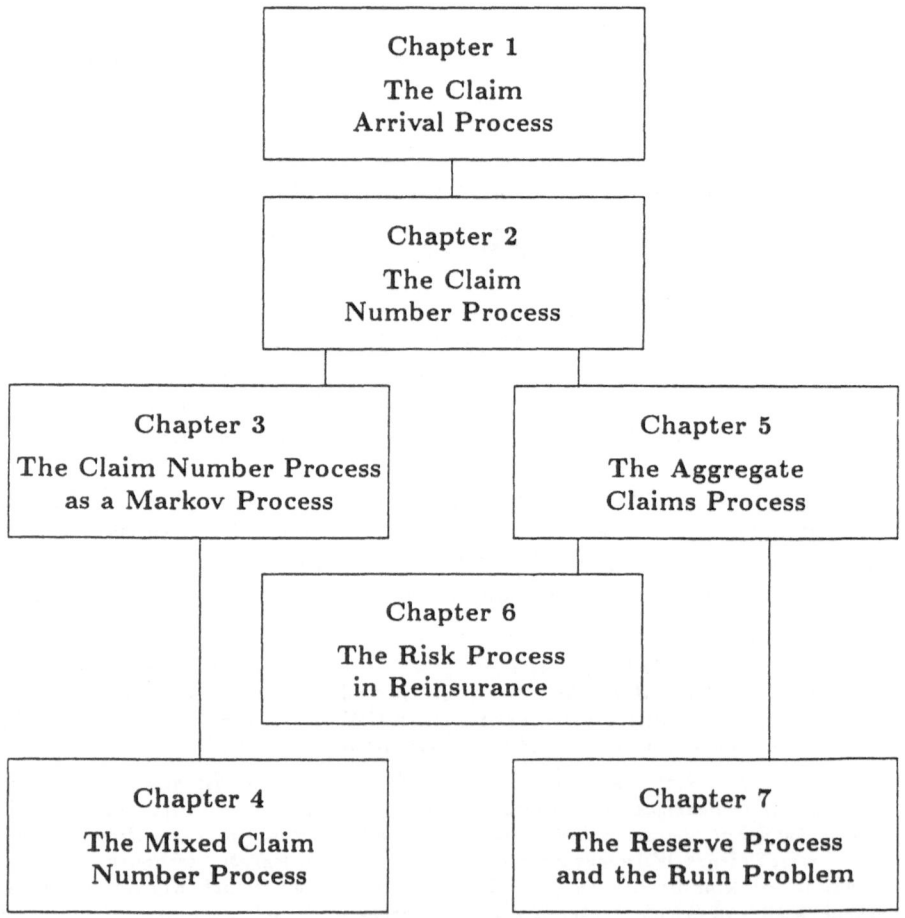

Interdependence Table

and then turn to the claim number process which will be studied in three chapters, exhibiting the properties of the Poisson process (Chapter 2) and of its extensions to Markov claim number processes (Chapter 3) and to mixed Poisson processes (Chapter 4). Mixed Poisson processes are particularly useful in applications since they reflect the idea of an inhomogeneous portfolio. We then pass to the aggregate claims process and study some methods of computing or estimating aggregate claims distributions (Chapter 5). A particular aggregate claims process is the thinned claim number process occurring in excess of loss reinsurance, where the reinsurer assumes responsibility for claim severities exceeding a given priority, and this leads to the discussion of thinning and the related problems of decomposition and superposition of risk processes (Chapter 6). Finally, we consider the reserve process and the ruin problem in an infinite time horizon when the premium income is proportional to time (Chapter 7).

The Interdependence Table given above indicates various possibilities for a selective reading of the book. For a first reading, it would be sufficient to study Chapters 1, 2, 5, and 7, but it should be noted that Chapter 2 is almost entirely devoted to the Poisson process. Since Poisson processes are unrealistic models in many classes of nonlife insurance, these chapters should be complemented by some of the material presented in Chapters 3, 4, and 6. A substantial part of Chapter 4 is independent of Chapter 3, and Chapters 6 and 7 contain only sporadic references to definitions or results of Chapter 3 and depend on Chapter 5 only via Section 5.1. Finally, a reader who is primarily interested in claim number processes may leave Chapter 5 after Section 5.1 and omit Chapter 7.

The reader of these notes is supposed to have a solid background in abstract measure and integration theory as well as some knowledge in measure theoretic probability theory; by contrast, particular experience with special distributions or stochastic processes is not required. All prerequisites can be found in the monographs by Aliprantis and Burkinshaw [1990], Bauer [1991, 1992], and Billingsley [1995].

Almost all proofs are given in great detail; some of them are elementary, others are rather involved, and certain proofs may seem to be superfluous since the result is suggested by the actuarial interpretation of the model. However, if actuarial mathematics is to be considered as a part of probability theory and mathematical statistics, then it has to accept its (sometimes bothering) rigour.

The notation in this book is standard, but for the convenience of the reader we fix some symbols and conventions; further details on notation may be found in the List of Symbols.

Throughout this book, let $(\Omega, \mathcal{F}, P)$ be a fixed probability space, let $\mathcal{B}(\mathbf{R}^n)$ denote the σ–algebra of Borel sets on the Euclidean space $\mathbf{R}^n$, let ξ denote the counting measure concentrated on $\mathbf{N}_0$, and let λ^n denote the Lebesgue measure on $\mathcal{B}(\mathbf{R}^n)$; in the case $n = 1$, the superscript n will be dropped.

The indicator function of a set A will be denoted by χ_A. A family of sets $\{A_i\}_{i\in I}$ is said to be *disjoint* if it is pairwise disjoint, and in this case its union will be denoted by $\sum_{i\in I} A_i$. A family of sets $\{A_i\}_{i\in I}$ is said to be a *partition* of A if it is disjoint and satisfies $\sum_{i\in I} A_i = A$.

For a sequence of random variables $\{Z_n\}_{n\in N}$ which are i. i. d. (independent and identically distributed), a typical random variable of the sequence will be denoted by Z. As a rule, integrals are Lebesgue integrals, but occasionally we have to switch to Riemann integrals in order to complete computations.

In some cases, sums, products, intersections, and unions extend over the empty index set; in this case, they are defined to be equal to zero, one, the reference set, and the empty set, respectively. The terms *positive, increasing*, etc. are used in the weak sense admitting equality.

The main concepts related to (probability) distributions as well as the definitions and the basic properties of the distributions referred to by name in this book are collected in the Appendix. Except for the Dirac distributions, all parametric families of distributions are defined as to exclude degenerate distributions and such that their parameters are taken from open intervals of $\mathbf{R}$ or subsets of $\mathbf{N}$.

It has been pointed out before that the present book addresses only a single topic of risk theory: The development in time of a fixed portfolio of risks. Other important topics in risk theory include the approximation of aggregate claims distributions, tariffication, reserving, and reinsurance, as well as the wide field of life insurance or, more generally, insurance of persons. The following references may serve as a guide to recent publications on some topics of actuarial mathematics which are beyond the scope of this book:
- *Life insurance mathematics*: Gerber [1986, 1990, 1995], Wolfsdorf [1986], Wolthuis [1994], and Helbig and Milbrodt [1995].
- *Life and nonlife insurance mathematics*: Bowers, Gerber, Hickman, Jones, and Nesbitt [1986], Panjer and Willmot [1992], and Daykin, Pentikäinen and Pesonen [1994].
- *Nonlife insurance mathematics*: Bühlmann [1970], Gerber [1979], Sundt [1984, 1991, 1993], Heilmann [1987, 1988], Straub [1988], Wolfsdorf [1988], Goovaerts, Kaas, van Heerwaarden, and Bauwelinckx [1990], Hipp and Michel [1990], and Norberg [1990].

Since the traditional distinction between life and nonlife insurance mathematics is becoming more and more obsolete, future research in actuarial mathematics should, in particular, aim at a unified theory providing models for all classes of insurance.

Chapter 1

The Claim Arrival Process

In order to model the development of an insurance business in time, we proceed in several steps by successively introducing
- the claim arrival process,
- the claim number process,
- the aggregate claims process, and
- the reserve process.
We shall see that claim arrival processes and claim number processes determine each other, and that claim number processes are the heart of the matter.

The present chapter is entirely devoted to the claim arrival process.

We first state the general model which will be studied troughout this book and which will be completed later (Section 1.1). We then study the special case of a claim arrival process having independent and identically exponentially distributed waiting times between two successive claims (Section 1.2). We finally show that the exponential distribution is of particular interest since it is the unique distribution which is memoryless on the interval $(0, \infty)$ (Section 1.3).

1.1 The Model

We consider a portfolio of risks which are insured by some insurer. The risks produce claims and pay premiums to the insurer who, in turn, will settle the claims. The portfolio may consist of a single risk or of several ones.

We assume that the insurer is primarily interested in the overall performance of the portfolio, that is, the balance of premiums and claim payments aggregated over all risks. (Of course, a surplus of premiums over claim payments would be welcome!) In the case where the portfolio consists of several risks, this means that the insurer does not care which of the risks in the portfolio causes a particular claim. This is the collective point of view in risk theory.

We assume further that in the portfolio claims occur at random in an infinite time horizon starting at time zero such that
– no claims occur at time zero, and
– no two claims occur simultaneously.
The assumption of no two claims occurring simultaneously seems to be harmless. Indeed, it should not present a serious problem when the portfolio is small; however, when the portfolio is large, it depends on the class of insurance under consideration whether this assumption is really acceptable. (For example, the situation is certainly different in fire insurance and in (third party liability) automobile insurance, where in certain countries a single insurance company holds about one quarter of all policies; in such a situation, one has to take into account the possibility that two insurees from the same large portfolio produce a car accident for which both are responsible in parts.)

Comment: When the assumption of no two claims occurring simultaneously is judged to be non–acceptable, it can nevertheless be saved by slightly changing the point of view, namely, by considering *claim events* (like car accidents) instead of single claims. The number of single claims occurring at a given claim event can then be interpreted as the size of the claim event. This point of view will be discussed further in Chapter 5 below.

Let us now transform the previous ideas into a probabilistic model:

A sequence of random variables $\{T_n\}_{n\in\mathbf{N}_0}$ is a *claim arrival process* if there exists a null set $\Omega_T \in \mathcal{F}$ such that, for all $\omega \in \Omega\backslash\Omega_T$,
– $T_0(\omega) = 0$ and
– $T_{n-1}(\omega) < T_n(\omega)$ holds for all $n \in \mathbf{N}$.
Then we have $T_n(\omega) > 0$ for all $n \in \mathbf{N}$ and all $\omega \in \Omega\backslash\Omega_T$. The null set Ω_T is said to be the *exceptional null set* of the claim arrival process $\{T_n\}_{n\in\mathbf{N}_0}$.

For a claim arrival process $\{T_n\}_{n\in\mathbf{N}_0}$ and for all $n \in \mathbf{N}$, define the increment

$$W_n \;:=\; T_n - T_{n-1} \;.$$

Then we have $W_n(\omega) > 0$ for all $n \in \mathbf{N}$ and all $\omega \in \Omega\backslash\Omega_T$, and hence

$$E[W_n] \;>\; 0$$

for all $n \in \mathbf{N}$, as well as

$$T_n \;=\; \sum_{k=1}^{n} W_k$$

for all $n \in \mathbf{N}$. The sequence $\{W_n\}_{n\in\mathbf{N}}$ is said to be the *claim interarrival process* induced by the claim arrival process $\{T_n\}_{n\in\mathbf{N}_0}$.

Interpretation:
- T_n is the *occurrence time* of the nth claim.
- W_n is the *waiting time* between the occurrence of claim $n-1$ and the occurrence of claim n.
- With probability one, no claim occurs at time zero and no two claims occur simultaneously.

For the remainder of this chapter, let $\{T_n\}_{n \in \mathbb{N}_0}$ be a fixed claim arrival process and let $\{W_n\}_{n \in \mathbb{N}}$ be the claim interarrival process induced by $\{T_n\}_{n \in \mathbb{N}_0}$. Without loss of generality, we may and do assume that the exceptional null set of the claim arrival process is empty.

Since $W_n = T_n - T_{n-1}$ and $T_n = \sum_{k=1}^{n} W_n$ holds for all $n \in \mathbb{N}$, it is clear that the claim arrival process and the claim interarrival process determine each other. In particular, we have the following obvious but useful result:

1.1.1 Lemma. *The identity*

$$\sigma\big(\{T_k\}_{k \in \{0,1,\dots,n\}}\big) \; = \; \sigma\big(\{W_k\}_{k \in \{1,\dots,n\}}\big)$$

holds for all $n \in \mathbb{N}$.

Furthermore, for $n \in \mathbb{N}$, let $\mathbf{T}_n$ and $\mathbf{W}_n$ denote the random vectors $\Omega \to \mathbb{R}^n$ with coordinates T_i and W_i, respectively, and let $\mathbf{M}_n$ denote the $(n \times n)$–matrix with entries

$$m_{ij} \; := \; \begin{cases} 1 & \text{if} \quad i \geq j \\ 0 & \text{if} \quad i < j \end{cases}.$$

Then $\mathbf{M}_n$ is invertible and satisfies $\det \mathbf{M}_n = 1$, and we have $\mathbf{T}_n = \mathbf{M}_n \circ \mathbf{W}_n$ and $\mathbf{W}_n = \mathbf{M}_n^{-1} \circ \mathbf{T}_n$. The following result is immediate:

1.1.2 Lemma. *For all $n \in \mathbb{N}$, the distributions of $\mathbf{T}_n$ and $\mathbf{W}_n$ satisfy*

$$P_{\mathbf{T}_n} \; = \; (P_{\mathbf{W}_n})_{\mathbf{M}_n} \quad and \quad P_{\mathbf{W}_n} \; = \; (P_{\mathbf{T}_n})_{\mathbf{M}_n^{-1}}.$$

The assumptions of our model do not exclude the possibility that infinitely many claims occur in finite time. The event

$$\{\sup_{n \in \mathbb{N}} T_n < \infty\}$$

is called *explosion*.

1.1.3 Lemma. *If $\sup_{n \in \mathbb{N}} E[T_n] < \infty$, then the probability of explosion is equal to one.*

This is obvious from the monotone convergence theorem.

1.1.4 Corollary.   *If $\sum_{n=1}^{\infty} E[W_n] < \infty$, then the probability of explosion is equal to one.*

In modelling a particular insurance business, one of the first decisions to take is to decide whether the probability of explosion should be zero or not. This decision is, of course, a decision concerning the distribution of the claim arrival process.

We conclude this section with a construction which in the following chapter will turn out to be a useful technical device:

For $n \in \mathbf{N}$, the graph of T_n is defined to be the map $U_n : \Omega \to \Omega \times \mathbf{R}$, given by

$$U_n(\omega) \; := \; \big(\omega, T_n(\omega)\big) \, .$$

Then each U_n is $\mathcal{F}$-$\mathcal{F} \otimes \mathcal{B}(\mathbf{R})$–measurable. Define a measure $\mu : \mathcal{F} \otimes \mathcal{B}(\mathbf{R}) \to [0, \infty]$ by letting

$$\mu[C] \; := \; \sum_{n=1}^{\infty} P_{U_n}[C] \, .$$

The measure μ will be called the *claim measure* induced by the claim arrival process $\{T_n\}_{n \in \mathbf{N}_0}$.

1.1.5 Lemma.   *The identity*

$$\mu[A \times B] \; = \; \int_A \left(\sum_{n=1}^{\infty} \chi_{\{T_n \in B\}} \right) dP$$

holds for all $A \in \mathcal{F}$ and $B \in \mathcal{B}(\mathbf{R})$.

Proof.   Since $U_n^{-1}(A \times B) = A \cap \{T_n \in B\}$, we have

$$\begin{aligned}
\mu[A \times B] \; &= \; \sum_{n=1}^{\infty} P_{U_n}[A \times B] \\
&= \; \sum_{n=1}^{\infty} P[A \cap \{T_n \in B\}] \\
&= \; \sum_{n=1}^{\infty} \int_A \chi_{\{T_n \in B\}} \, dP \\
&= \; \int_A \left(\sum_{n=1}^{\infty} \chi_{\{T_n \in B\}} \right) dP \, ,
\end{aligned}$$

as was to be shown.   □

The previous result connects the claim measure with the claim number process which will be introduced in Chapter 2.

Most results in this book involving special distributions concern the case where the distributions of the claim arrival times are absolutely continuous with respect to Lebesgue measure; this case will be referred to as the *continuous time model*. It is, however, quite interesting to compare the results for the continuous time model with corresponding ones for the case where the distributions of the claim arrival times are absolutely continuous with respect to the counting measure concentrated on N_0. In the latter case, there is no loss of generality if we assume that the claim arrival times are integer–valued, and this case will be referred to as the *discrete time model*. The discrete time model is sometimes considered to be an approximation of the continuous time model if the time unit is small, but we shall see that the properties of the discrete time model may drastically differ from those of the continuous time model. On the other hand, the discrete time model may also serve as a simple model in its own right if the portfolio is small and if the insurer merely wishes to distinguish claim–free periods from periods with a strictly positive number of claims. Results for the discrete time model will be stated as problems in this and subsequent chapters.

Another topic which is related to our model is *life insurance*. In the simplest case, we consider a single random variable T satisfying $P[\{T > 0\}] = 1$, which is interpreted as the *time of death* or the *lifetime* of the insured individual; accordingly, this model is called *single life insurance*. More generally, we consider a finite sequence of random variables $\{T_n\}_{n \in \{0,1,...,N\}}$ satisfying $P[\{T_0 = 0\}] = 1$ and $P[\{T_{n-1} < T_n\}] = 1$ for all $n \in \{1, \ldots, N\}$, where T_n is interpreted as the *time of the nth death* in a portfolio of N insured individuals; accordingly, this model is called *multiple life insurance*. Although life insurance will not be studied in detail in these notes, some aspects of single or multiple life insurance will be discussed as problems in this and subsequent chapters.

Problems

1.1.A   If the sequence of claim interarrival times is i. i. d., then the probability of explosion is equal to zero.

1.1.B   **Discrete Time Model:** The inequality $T_n \geq n$ holds for all $n \in N$.

1.1.C   **Discrete Time Model:** The probability of explosion is equal to zero.

1.1.D   **Multiple Life Insurance:** Extend the definition of a claim arrival process as to cover the case of multiple (and hence single) life insurance.

1.1.E   **Multiple Life Insurance:** The probability of explosion is equal to zero.

1.2   The Erlang Case

In some of the special cases of our model which we shall discuss in detail, the claim interarrival times are assumed or turn out to be independent and exponentially distributed. In this situation, explosion is either impossible or certain:

1.2.1 Theorem (Zero–One Law on Explosion). *Let $\{\alpha_n\}_{n\in\mathbf{N}}$ be a sequence of real numbers in $(0,\infty)$ and assume that the sequence of claim interarrival times $\{W_n\}_{n\in\mathbf{N}}$ is independent and satisfies $P_{W_n} = \mathbf{Exp}(\alpha_n)$ for all $n\in\mathbf{N}$.*
(a) *If the series $\sum_{n=1}^{\infty} 1/\alpha_n$ diverges, then the probability of explosion is equal to zero.*
(b) *If the series $\sum_{n=1}^{\infty} 1/\alpha_n$ converges, then the probability of explosion is equal to one.*

Proof. By the dominated convergence theorem, we have

$$
E\left[e^{-\sum_{n=1}^{\infty} W_n}\right] = E\left[\prod_{n=1}^{\infty} e^{-W_n}\right]
$$
$$
= \prod_{n=1}^{\infty} E\left[e^{-W_n}\right]
$$
$$
= \prod_{n=1}^{\infty} \frac{\alpha_n}{\alpha_n + 1}
$$
$$
= \prod_{n=1}^{\infty} \left(1 - \frac{1}{1+\alpha_n}\right)
$$
$$
\leq \prod_{n=1}^{\infty} e^{-1/(1+\alpha_n)}
$$
$$
= e^{-\sum_{n=1}^{\infty} 1/(1+\alpha_n)} \, .
$$

Thus, if the series $\sum_{n=1}^{\infty} 1/\alpha_n$ diverges, then the series $\sum_{n=1}^{\infty} 1/(1+\alpha_n)$ diverges as well and we have $P[\{\sum_{n=1}^{\infty} W_n = \infty\}] = 1$, and thus

$$
P[\{\sup_{n\in\mathbf{N}} T_n < \infty\}] = P\left[\left\{\sum_{n=1}^{\infty} W_n < \infty\right\}\right]
$$
$$
= 0 \, ,
$$

which proves (a).
Assertion (b) is immediate from Corollary 1.1.4. $\square$

In the case of independent claim interarrival times, the following result is also of interest:

1.2.2 Lemma. *Let $\alpha \in (0,\infty)$. If the sequence of claim interarrival times $\{W_n\}_{n\in\mathbf{N}}$ is independent, then the following are equivalent:*
(a) *$P_{W_n} = \mathbf{Exp}(\alpha)$ for all $n\in\mathbf{N}$.*
(b) *$P_{T_n} = \mathbf{Ga}(\alpha,n)$ for all $n\in\mathbf{N}$.*
In this case, $E[W_n] = 1/\alpha$ and $E[T_n] = n/\alpha$ holds for all $n \in \mathbf{N}$, and the probability of explosion is equal to zero.

Proof. The simplest way to prove the equivalence of (a) and (b) is to use characteristic functions.

• Assume first that (a) holds. Since $T_n = \sum_{k=1}^{n} W_k$, we have

$$\varphi_{T_n}(z) = \prod_{k=1}^{n} \varphi_{W_k}(z)$$

$$= \prod_{k=1}^{n} \frac{\alpha}{\alpha - iz}$$

$$= \left(\frac{\alpha}{\alpha - iz} \right)^n ,$$

and thus $P_{T_n} = \mathbf{Ga}(\alpha, n)$. Therefore, (a) implies (b).

• Assume now that (b) holds. Since $T_{n-1} + W_n = T_n$, we have

$$\left(\frac{\alpha}{\alpha - iz} \right)^{n-1} \cdot \varphi_{W_n}(z) = \varphi_{T_{n-1}}(z) \cdot \varphi_{W_n}(z)$$

$$= \varphi_{T_n}(z)$$

$$= \left(\frac{\alpha}{\alpha - iz} \right)^n ,$$

hence

$$\varphi_{W_n}(z) = \frac{\alpha}{\alpha - iz} ,$$

and thus $P_{W_n} = \mathbf{Exp}(\alpha)$. Therefore, (b) implies (a).

• The final assertion is obvious from the distributional assumptions and the zero–one law on explosion.

• For readers not familiar with characteristic functions, we include an elementary proof of the implication (a) $\implies$ (b); only this implication will be needed in the sequel. Assume that (a) holds. We proceed by induction.

Obviously, since $T_1 = W_1$ and $\mathbf{Exp}(\alpha) = \mathbf{Ga}(\alpha, 1)$, we have $P_{T_1} = \mathbf{Ga}(\alpha, 1)$.

Assume now that $P_{T_n} = \mathbf{Ga}(\alpha, n)$ holds for some $n \in \mathbf{N}$. Then we have

$$P_{T_n}[B] = \int_B \frac{\alpha^n}{\Gamma(n)} e^{-\alpha x} x^{n-1} \chi_{(0,\infty)}(x) \, d\lambda(x)$$

and

$$P_{W_{n+1}}[B] = \int_B \alpha e^{-\alpha x} \chi_{(0,\infty)}(x) \, d\lambda(x)$$

for all $B \in \mathcal{B}(\mathbf{R})$. Since T_n and W_{n+1} are independent, the convolution formula yields

$$P_{T_{n+1}}[B] = P_{T_n + W_{n+1}}[B]$$

$$= P_{T_n} * P_{W_{n+1}}[B]$$

$$= \int_B \left(\int_{\mathbf{R}} \frac{\alpha^n}{\Gamma(n)} e^{-\alpha(t-s)} (t-s)^{n-1} \chi_{(0,\infty)}(t-s) \, \alpha e^{-\alpha s} \chi_{(0,\infty)}(s) \, d\lambda(s) \right) d\lambda(t)$$

$$= \int_B \frac{\alpha^{n+1}}{\Gamma(n+1)} e^{-\alpha t} \left(\int_{\mathbf{R}} n(t-s)^{n-1} \chi_{(0,t)}(s) \, d\lambda(s) \right) \chi_{(0,\infty)}(t) \, d\lambda(t)$$

$$= \int_B \frac{\alpha^{n+1}}{\Gamma(n+1)} e^{-\alpha t} \left(\int_0^t n(t-s)^{n-1} \, ds \right) \chi_{(0,\infty)}(t) \, d\lambda(t)$$

$$= \int_B \frac{\alpha^{n+1}}{\Gamma(n+1)} e^{-\alpha t} \, t^n \, \chi_{(0,\infty)}(t) \, d\lambda(t)$$

$$= \int_B \frac{\alpha^{n+1}}{\Gamma(n+1)} e^{-\alpha t} \, t^{(n+1)-1} \chi_{(0,\infty)}(t) \, d\lambda(t)$$

for all $B \in \mathcal{B}(\mathbf{R})$, and thus $P_{T_{n+1}} = \mathbf{Ga}(\alpha, n+1)$. Therefore, (a) implies (b). $\square$

The particular role of the exponential distribution will be discussed in the following section.

Problems

1.2.A   Let $Q := \mathbf{Exp}(\alpha)$ for some $\alpha \in (0, \infty)$ and let Q' denote the unique distribution satisfying $Q'[\{k\}] = Q[(k-1, k]]$ for all $k \in \mathbf{N}$. Then $Q' = \mathbf{Geo}(1 - e^{-\alpha})$.

1.2.B   **Discrete Time Model:** Let $\vartheta \in (0, 1)$. If the sequence $\{W_n\}_{n \in \mathbf{N}}$ is independent, then the following are equivalent:
(a) $P_{W_n} = \mathbf{Geo}(\vartheta)$ for all $n \in \mathbf{N}$.
(b) $P_{T_n} = \mathbf{Geo}(n, \vartheta)$ for all $n \in \mathbf{N}$.
In this case, $E[W_n] = 1/\vartheta$ and $E[T_n] = n/\vartheta$ holds for all $n \in \mathbf{N}$.

1.3   A Characterization of the Exponential Distribution

One of the most delicate problems in probabilistic modelling is the appropriate choice of the distributions of the random variables in the model. More precisely, it is the joint distribution of all random variables that has to be specified. To make this choice, it is useful to know that certain distributions are characterized by general properties which are easy to interpret.

In the model considered here, it is sufficient to specify the distribution of the claim interarrival process. This problem is considerably reduced if the claim interarrival times are assumed to be independent, but even in that case the appropriate choice of the distributions of the single claim interarrival times is not obvious. In what follows we shall characterize the exponential distribution by a simple property which is helpful to decide whether or not in a particular insurance business this distribution is appropriate for the claim interarrival times.

For the moment, consider a random variable W which may be interpreted as a waiting time.

If $P_W = \mathbf{Exp}(\alpha)$, then the *survival function* $\mathbf{R} \to [0,1] : w \mapsto P[\{W > w\}]$ of the distribution of W satisfies $P[\{W > w\}] = e^{-\alpha w}$ for all $w \in \mathbf{R}_+$, and this yields

$$P[\{W > s+t\}] \;\; = \;\; P[\{W > s\}] \cdot P[\{W > t\}]$$

or, equivalently,

$$P[\{W > s+t\} | \{W > s\}] \;\; = \;\; P[\{W > t\}]$$

for all $s, t \in \mathbf{R}_+$. The first identity reflects the fact that the survival function of the exponential distribution is self–similar on $\mathbf{R}_+$ in the sense that, for each $s \in \mathbf{R}_+$, the graphs of the mappings $t \mapsto P[\{W > s+t\}]$ and $t \mapsto P[\{W > t\}]$ differ only by a scaling factor. Moreover, if W is interpreted as a waiting time, then the second identity means that the knowledge of having waited more than s time units does not provide any information on the remaining waiting time. Loosely speaking, the exponential distribution has no memory (or does not use it). The question arises whether the exponential distribution is the unique distribution having this property.

Before formalizing the notion of a memoryless distribution, we observe that in the case $P_W = \mathbf{Exp}(\alpha)$ the above identities hold for all $s, t \in \mathbf{R}_+$ but fail for all $s, t \in \mathbf{R}$ such that $s < 0 < s+t$; on the other hand, we have $P_W[\mathbf{R}_+] = 1$. These observations lead to the following definition:

A distribution $Q : \mathcal{B}(\mathbf{R}) \to [0,1]$ is *memoryless on* $S \in \mathcal{B}(\mathbf{R})$ if
- $Q[S] = 1$ and
- the identity

$$Q[(s+t, \infty)] \;\; = \;\; Q[(s, \infty)] \cdot Q[(t, \infty)]$$

holds for all $s, t \in S$.

The following result yields a general property of memoryless distributions:

1.3.1 Theorem. *Let* $Q : \mathcal{B}(\mathbf{R}) \to [0,1]$ *be a distribution which is memoryless on* $S \in \mathcal{B}(\mathbf{R})$. *If* $0 \in S$, *then* Q *satisfies either* $Q[\{0\}] = 1$ *or* $Q[(0, \infty)] = 1$.

Proof. Assume that $Q[(0, \infty)] < 1$. Since $0 \in S$, we have

$$
\begin{aligned}
Q[(0, \infty)] \;\; &= \;\; Q[(0, \infty)] \cdot Q[(0, \infty)] \\
&= \;\; Q[(0, \infty)]^2 \, ,
\end{aligned}
$$

hence

$$Q[(0, \infty)] \;=\; 0 \,,$$

and thus

$$\begin{aligned} Q[(t, \infty)] &= Q[(t, \infty)] \cdot Q[(0, \infty)] \\ &= 0 \end{aligned}$$

for all $t \in S$.

Define $t := \inf S$ and choose a sequence $\{t_n\}_{n \in \mathbf{N}} \subseteq S$ which decreases to t. Then we have

$$\begin{aligned} Q[(t, \infty)] &= \sup_{n \in \mathbf{N}} Q[(t_n, \infty)] \\ &= 0 \,. \end{aligned}$$

Since $Q[S] = 1$, we also have

$$Q[(-\infty, t)] \;=\; 0 \,.$$

Therefore, we have

$$Q[\{t\}] \;=\; 1 \,,$$

hence $Q[\{t\} \cap S] = 1$, and thus $t \in S$.

Finally, since $0 \in S$, we have either $t < 0$ or $t = 0$. But $t < 0$ implies $t \in (2t, \infty)$ and hence

$$\begin{aligned} Q[\{t\}] &\leq Q[(2t, \infty)] \\ &= Q[(t, \infty)] \cdot Q[(t, \infty)] \\ &= Q[(t, \infty)]^2 \,, \end{aligned}$$

which is impossible. Therefore, we have $t = 0$ and hence $Q[\{0\}] = 1$, as was to be shown. $\square$

The following result characterizes the exponential distribution:

1.3.2 Theorem. *For a distribution $Q : \mathcal{B}(\mathbf{R}) \to [0, 1]$, the following are equivalent:*
(a) *Q is memoryless on $(0, \infty)$.*
(b) *$Q = \mathbf{Exp}(\alpha)$ for some $\alpha \in (0, \infty)$.*
In this case, $\alpha = - \log Q[(1, \infty)]$.

Proof. Note that $Q = \mathbf{Exp}(\alpha)$ if and only if the identity

$$Q[(t, \infty)] \;=\; e^{-\alpha t}$$

holds for all $t \in [0, \infty)$.

• Assume that (a) holds. By induction, we have

$$Q[(n, \infty)] \;=\; Q[(1, \infty)]^n$$

and

$$Q[(1, \infty)] \;=\; Q[(1/n, \infty)]^n$$

for all $n \in \mathbf{N}$.

Thus, $Q[(1, \infty)] = 1$ is impossible because of

$$
\begin{aligned}
0 \;&=\; Q[\emptyset] \\
&=\; \inf_{n \in \mathbf{N}} Q[(n, \infty)] \\
&=\; \inf_{n \in \mathbf{N}} Q[(1, \infty)]^n \,,
\end{aligned}
$$

and $Q[(1, \infty)] = 0$ is impossible because of

$$
\begin{aligned}
1 \;&=\; Q[(0, \infty)] \\
&=\; \sup_{n \in \mathbf{N}} Q[(1/n, \infty)] \\
&=\; \sup_{n \in \mathbf{N}} Q[(1, \infty)]^{1/n} \,.
\end{aligned}
$$

Therefore, we have

$$Q[(1, \infty)] \;\in\; (0, 1) \,.$$

Define now $\alpha := -\log Q[(1, \infty)]$. Then we have $\alpha \in (0, \infty)$ and

$$Q[(1, \infty)] \;=\; e^{-\alpha} \,,$$

and thus

$$
\begin{aligned}
Q[(m/n, \infty)] \;&=\; Q[(1, \infty)]^{m/n} \\
&=\; \left(e^{-\alpha}\right)^{m/n} \\
&=\; e^{-\alpha m/n}
\end{aligned}
$$

for all $m, n \in \mathbf{N}$. This yields

$$Q[(t, \infty)] \;=\; e^{-\alpha t}$$

for all $t \in (0, \infty) \cap \mathbf{Q}$. Finally, for each $t \in [0, \infty)$ we may choose a sequence $\{t_n\}_{n \in \mathbf{N}} \subseteq (0, \infty) \cap \mathbf{Q}$ which decreases to t, and we obtain

$$
\begin{aligned}
Q[(t, \infty)] \;&=\; \sup_{n \in \mathbf{N}} Q[(t_n, \infty)] \\
&=\; \sup_{n \in \mathbf{N}} e^{-\alpha t_n} \\
&=\; e^{-\alpha t} \,.
\end{aligned}
$$

By the introductory remark, it follows that $Q = \mathbf{Exp}(\alpha)$. Therefore, (a) implies (b).

• The converse implication is obvious.   □

1.3.3 Corollary. *For a distribution $Q : \mathcal{B}(\mathbf{R}) \to [0,1]$, the following are equivalent:*
(a) *Q is memoryless on $\mathbf{R}_+$.*
(b) *Either $Q = \delta_0$ or $Q = \mathbf{Exp}(\alpha)$ for some $\alpha \in (0, \infty)$.*

Proof. The assertion is immediate from Theorems 1.3.1 and 1.3.2.   □

With regard to the previous result, note that the Dirac distribution δ_0 is the limit of the exponential distributions $\mathbf{Exp}(\alpha)$ as $\alpha \to \infty$.

1.3.4 Corollary. *There is no distribution which is memoryless on $\mathbf{R}$.*

Proof. If $Q : \mathcal{B}(\mathbf{R}) \to [0,1]$ is a distribution which is memoryless on $\mathbf{R}$, then either $Q = \delta_0$ or $Q = \mathbf{Exp}(\alpha)$ for some $\alpha \in (0, \infty)$, by Theorem 1.3.1 and Corollary 1.3.3. On the other hand, none of these distributions is memoryless on $\mathbf{R}$.   □

Problems

1.3.A   **Discrete Time Model:** For a distribution $Q : \mathcal{B}(\mathbf{R}) \to [0,1]$, the following are equivalent:
       (a) Q is memoryless on $\mathbf{N}$.
       (b) Either $Q = \delta_1$ or $Q = \mathbf{Geo}(\vartheta)$ for some $\vartheta \in (0,1)$.
       Note that the Dirac distribution δ_1 is the limit of the geometric distributions $\mathbf{Geo}(\vartheta)$ as $\vartheta \to 0$.

1.3.B   **Discrete Time Model:** For a distribution $Q : \mathcal{B}(\mathbf{R}) \to [0,1]$, the following are equivalent:
       (a) Q is memoryless on $\mathbf{N}_0$.
       (b) Either $Q = \delta_0$ or $Q = \delta_1$ or $Q = \mathbf{Geo}(\vartheta)$ for some $\vartheta \in (0,1)$.
       In particular, the negativebinomial distribution fails to be memoryless on $\mathbf{N}_0$.

1.3.C   There is no distribution which is memoryless on $(-\infty, 0)$.

1.4   Remarks

Since the conclusions obtained in a probabilistic model usually concern probabilities and not single realizations of random variables, it is natural to state the assumptions of the model in terms of probabilities as well. While this is a merely formal justification for the exceptional null set in the definition of the claim arrival process, there is also a more substantial reason: As we shall see in Chapters 5 and 6 below, it is sometimes of interest to construct a claim arrival process from other random variables, and in that case it cannot in general be ensured that the exceptional null set is empty.

Theorem 1.3.2 is the most famous characterization of the exponential distribution. Further characterizations of the exponential distribution can be found in the monographs by Galambos and Kotz [1978] and, in particular, by Azlarov and Volodin [1986].

Chapter 2

The Claim Number Process

In the previous chapter, we have formulated a general model for the occurrence of claims in an insurance business and we have studied the claim arrival process in some detail.

In the present chapter, we proceed one step further by introducing the claim number process. Particular attention will be given to the Poisson process.

We first introduce the general claim number process and show that claim number processes and claim arrival processes determine each other (Section 2.1). We then establish a connection between certain assumptions concerning the distributions of the claim arrival times and the distributions of the claim numbers (Section 2.2). We finally prove the main result of this chapter which characterizes the (homogeneous) Poisson process in terms of the claim interarrival process, the claim measure, and a martingale property (Section 2.3).

2.1 The Model

A family of random variables $\{N_t\}_{t \in \mathbf{R}_+}$ is a *claim number process* if there exists a null set $\Omega_N \in \mathcal{F}$ such that, for all $\omega \in \Omega \backslash \Omega_N$,

- $N_0(\omega) = 0$,
- $N_t(\omega) \in \mathbf{N}_0 \cup \{\infty\}$ for all $t \in (0, \infty)$,
- $N_t(\omega) = \inf_{s \in (t, \infty)} N_s(\omega)$ for all $t \in \mathbf{R}_+$,
- $\sup_{s \in [0, t)} N_s(\omega) \leq N_t(\omega) \leq \sup_{s \in [0, t)} N_s(\omega) + 1$ for all $t \in \mathbf{R}_+$, and
- $\sup_{t \in \mathbf{R}_+} N_t(\omega) = \infty$.

The null set Ω_N is said to be the *exceptional null set* of the claim number process $\{N_t\}_{t \in \mathbf{R}_+}$.

Interpretation:
- N_t is the *number of claims* occurring in the interval $(0, t]$.
- Almost all paths of $\{N_t\}_{t \in \mathbf{R}_+}$ start at zero, are right–continuous, increase with jumps of height one at discontinuity points, and increase to infinity.

Our first result asserts that every claim arrival process induces a claim number process, and vice versa:

2.1.1 Theorem.
(a) *Let $\{T_n\}_{n \in \mathbf{N}_0}$ be a claim arrival process. For all $t \in \mathbf{R}_+$ and $\omega \in \Omega$, define*

$$N_t(\omega) := \sum_{n=1}^{\infty} \chi_{\{T_n \leq t\}}(\omega) .$$

Then $\{N_t\}_{t \in \mathbf{R}_+}$ is a claim number process such that $\Omega_N = \Omega_T$, and the identity

$$T_n(\omega) = \inf\{t \in \mathbf{R}_+ \mid N_t(\omega) = n\}$$

holds for all $n \in \mathbf{N}_0$ and all $\omega \in \Omega \backslash \Omega_T$.
(b) *Let $\{N_t\}_{t \in \mathbf{R}_+}$ be a claim number process. For all $n \in \mathbf{N}_0$ and $\omega \in \Omega$, define*

$$T_n(\omega) := \inf\{t \in \mathbf{R}_+ \mid N_t(\omega) = n\} .$$

Then $\{T_n\}_{n \in \mathbf{N}_0}$ is a claim arrival process such that $\Omega_T = \Omega_N$, and the identity

$$N_t(\omega) = \sum_{n=1}^{\infty} \chi_{\{T_n \leq t\}}(\omega)$$

holds for all $t \in \mathbf{R}_+$ and all $\omega \in \Omega \backslash \Omega_N$.

The verification of Theorem 2.1.1 is straightforward.

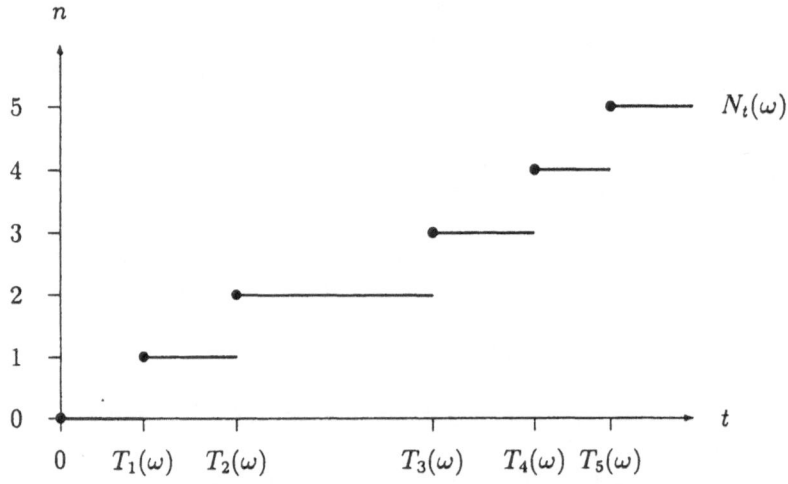

Claim Arrival Process and Claim Number Process

For the remainder of this chapter, let $\{N_t\}_{t \in \mathbf{R}_+}$ be a claim number process, let $\{T_n\}_{n \in \mathbf{N}_0}$ be the claim arrival process induced by the claim number process, and let $\{W_n\}_{n \in \mathbf{N}}$ be the claim interarrival process induced by the claim arrival process. We assume that the exceptional null set is empty.

By virtue of the assumption that the exceptional null set is empty, we have two simple but most useful identities, showing that certain events determined by the claim number process can be interpreted as events determined by the claim arrival process, and vice versa:

2.1.2 Lemma. *The identities*
(a) $\{N_t \geq n\} = \{T_n \leq t\}$ *and*
(b) $\{N_t = n\} = \{T_n \leq t\} \backslash \{T_{n+1} \leq t\} = \{T_n \leq t < T_{n+1}\}$
hold for all $n \in \mathbf{N}_0$ *and* $t \in \mathbf{R}_+$.

The following result expresses in a particularly concise way the fact that the claim number process and the claim arrival process contain the same information:

2.1.3 Lemma.
$$\sigma(\{N_t\}_{t \in \mathbf{R}_+}) = \sigma(\{T_n\}_{n \in \mathbf{N}_0}).$$

In view of the preceding discussion, it is not surprising that explosion can also be expressed in terms of the claim number process:

2.1.4 Lemma. *The probability of explosion satisfies*
$$P[\{\sup_{n \in \mathbf{N}} T_n < \infty\}] = P\left[\bigcup_{t \in \mathbf{N}} \{N_t = \infty\}\right] = P\left[\bigcup_{t \in (0,\infty)} \{N_t = \infty\}\right].$$

Proof. Since the family of sets $\{\{N_t = \infty\}\}_{t \in (0,\infty)}$ is increasing, we have
$$\bigcup_{t \in (0,\infty)} \{N_t = \infty\} = \bigcup_{t \in \mathbf{N}} \{N_t = \infty\}.$$

By Lemma 2.1.2, this yields
$$\begin{aligned}
\bigcup_{t \in (0,\infty)} \{N_t = \infty\} &= \bigcup_{t \in \mathbf{N}} \{N_t = \infty\} \\
&= \bigcup_{t \in \mathbf{N}} \bigcap_{n \in \mathbf{N}} \{N_t \geq n\} \\
&= \bigcup_{t \in \mathbf{N}} \bigcap_{n \in \mathbf{N}} \{T_n \leq t\} \\
&= \bigcup_{t \in \mathbf{N}} \{\sup_{n \in \mathbf{N}} T_n \leq t\} \\
&= \{\sup_{n \in \mathbf{N}} T_n < \infty\},
\end{aligned}$$

and the assertion follows.   □

2.1.5 Corollary. *Assume that the claim number process has finite expectations. Then the probability of explosion is equal to zero.*

Proof. By assumption, we have $E[N_t] < \infty$ and hence $P[\{N_t = \infty\}] = 0$ for all $t \in (0,\infty)$. The assertion now follows from Lemma 2.1.4.   □

The discussion of the claim number process will to a considerable extent rely on the properties of its increments which are defined as follows:

For $s, t \in \mathbf{R}_+$ such that $s \leq t$, the *increment* of the claim number process $\{N_t\}_{t \in \mathbf{R}_+}$ on the interval $(s, t]$ is defined to be

$$N_t - N_s := \sum_{n=1}^{\infty} \chi_{\{s < T_n \leq t\}} .$$

Since $N_0 = 0$ and $T_n > 0$ for all $n \in \mathbf{N}$, this is in accordance with the definition of N_t; in addition, we have

$$N_t(\omega) = (N_t - N_s)(\omega) + N_s(\omega) ,$$

even if $N_s(\omega)$ is infinite.

The final result in this section connects the increments of the claim number process, and hence the claim number process itself, with the claim measure:

2.1.6 Lemma. *The identity*

$$\mu[A \times (s, t]] = \int_A (N_t - N_s) \, dP$$

holds for all $A \in \mathcal{F}$ and $s, t \in \mathbf{R}_+$ such that $s \leq t$.

Proof. The assertion follows from Lemma 1.1.5 and the definition of $N_t - N_s$. $\quad\square$

In the discrete time model, we have $N_t = N_{t+h}$ for all $t \in \mathbf{N}_0$ and $h \in [0, 1)$ so that nothing is lost if in this case the index set of the claim number process $\{N_t\}_{t \in \mathbf{R}_+}$ is reduced to $\mathbf{N}_0$; we shall then refer to the sequence $\{N_l\}_{l \in \mathbf{N}_0}$ as the claim number process induced by $\{T_n\}_{n \in \mathbf{N}_0}$.

Problems

2.1.A   **Discrete Time Model:** The inequalities
     (a) $N_l \leq N_{l-1} + 1$ and
     (b) $N_l \leq l$
     hold for all $l \in \mathbf{N}$.

2.1.B   **Discrete Time Model:** The identities
     (a) $\{N_l - N_{l-1} = 0\} = \sum_{j=1}^{l} \{T_{j-1} < l < T_j\}$,
         $\{N_l - N_{l-1} = 1\} = \sum_{j=1}^{l} \{T_j = l\}$, and
     (b) $\{T_n = l\} = \{n \leq N_l\} \backslash \{n \leq N_{l-1}\} = \{N_{l-1} < n \leq N_l\}$
     hold for all $n \in \mathbf{N}$ and $l \in \mathbf{N}$.

2.1.C **Multiple Life Insurance:** For all $t \in \mathbf{R}_+$, define $N_t := \sum_{n=1}^{N} \chi_{\{T_n \leq t\}}$. Then there exists a null set $\Omega_N \in \mathcal{F}$ such that, for all $\omega \in \Omega \setminus \Omega_N$,
  – $N_0(\omega) = 0$,
  – $N_t(\omega) \in \{0, 1, \ldots, N\}$ for all $t \in (0, \infty)$,
  – $N_t(\omega) = \inf_{s \in (t, \infty)} N_s(\omega)$ for all $t \in \mathbf{R}_+$,
  – $\sup_{s \in [0,t)} N_s(\omega) \leq N_t(\omega) \leq \sup_{s \in [0,t)} N_s(\omega) + 1$ for all $t \in \mathbf{R}_+$, and
  – $\sup_{t \in \mathbf{R}_+} N_t(\omega) = N$.

2.2 The Erlang Case

In the present section we return to the special case of claim arrival times having an Erlang distribution.

2.2.1 Lemma. *Let $\alpha \in (0, \infty)$. Then the following are equivalent:*
(a) $P_{T_n} = \mathbf{Ga}(\alpha, n)$ *for all $n \in \mathbf{N}$.*
(b) $P_{N_t} = \mathbf{P}(\alpha t)$ *for all $t \in (0, \infty)$.*
In this case, $E[T_n] = n/\alpha$ holds for all $n \in \mathbf{N}$ and $E[N_t] = \alpha t$ holds for all $t \in (0, \infty)$.

Proof. Note that the identity

$$e^{-\alpha t} \frac{(\alpha t)^n}{n!} = \int_0^t \frac{\alpha^n}{\Gamma(n)} e^{-\alpha s} s^{n-1} \, ds - \int_0^t \frac{\alpha^{n+1}}{\Gamma(n+1)} e^{-\alpha s} s^n \, ds$$

holds for all $n \in \mathbf{N}$ and $t \in (0, \infty)$.

• Assume first that (a) holds. Lemma 2.1.2 yields

$$\begin{aligned} P[\{N_t = 0\}] &= P[\{t < T_1\}] \\ &= e^{-\alpha t}, \end{aligned}$$

and, for all $n \in \mathbf{N}$,

$$\begin{aligned} P[\{N_t = n\}] &= P[\{T_n \leq t\}] - P[\{T_{n+1} \leq t\}] \\ &= \int_{(-\infty, t]} \frac{\alpha^n}{\Gamma(n)} e^{-\alpha s} s^{n-1} \chi_{(0,\infty)}(s) \, d\lambda(s) \\ &\quad - \int_{(-\infty, t]} \frac{\alpha^{n+1}}{\Gamma(n+1)} e^{-\alpha s} s^{(n+1)-1} \chi_{(0,\infty)}(s) \, d\lambda(s) \\ &= \int_0^t \frac{\alpha^n}{\Gamma(n)} e^{-\alpha s} s^{n-1} \, ds - \int_0^t \frac{\alpha^{n+1}}{\Gamma(n+1)} e^{-\alpha s} s^n \, ds \\ &= e^{-\alpha t} \frac{(\alpha t)^n}{n!}. \end{aligned}$$

This yields

$$P[\{N_t = n\}] = e^{-\alpha t} \frac{(\alpha t)^n}{n!}$$

for all $n \in \mathbf{N}_0$, and hence $P_{N_t} = \mathbf{P}(\alpha t)$. Therefore, (a) implies (b).

- Assume now that (b) holds. Since $T_n > 0$, we have

$$P[\{T_n \leq t\}] \ = \ 0$$

for all $t \in (-\infty, 0]$; also, for all $t \in (0, \infty)$, Lemma 2.1.2 yields

$$
\begin{aligned}
P[\{T_n \leq t\}] \ &= \ P[\{N_t \geq n\}] \\
&= \ 1 - P[\{N_t \leq n-1\}] \\
&= \ 1 - \sum_{k=0}^{n-1} P[\{N_t = k\}] \\
&= \ 1 - \sum_{k=0}^{n-1} e^{-\alpha t} \frac{(\alpha t)^k}{k!} \\
&= \ (1 - e^{-\alpha t}) - \sum_{k=1}^{n-1} e^{-\alpha t} \frac{(\alpha t)^k}{k!} \\
&= \ \int_0^t \alpha e^{-\alpha s}\, ds - \sum_{k=1}^{n-1} \left(\int_0^t \frac{\alpha^k}{\Gamma(k)} e^{-\alpha s} s^{k-1}\, ds - \int_0^t \frac{\alpha^{k+1}}{\Gamma(k+1)} e^{-\alpha s} s^k\, ds \right) \\
&= \ \int_0^t \frac{\alpha^n}{\Gamma(n)} e^{-\alpha s} s^{n-1}\, ds \\
&= \ \int_{(-\infty, t]} \frac{\alpha^n}{\Gamma(n)} e^{-\alpha s} s^{n-1} \chi_{(0,\infty)}(s)\, d\lambda(s) \ .
\end{aligned}
$$

This yields

$$P[\{T_n \leq t\}] \ = \ \int_{(-\infty, t]} \frac{\alpha^n}{\Gamma(n)} e^{-\alpha s} s^{n-1} \chi_{(0,\infty)}(s)\, d\lambda(s)$$

for all $t \in \mathbf{R}$, and hence $P_{T_n} = \mathbf{Ga}(\alpha, n)$. Therefore, (b) implies (a).
- The final assertion is obvious.   □

By Lemma 1.2.2, the equivalent conditions of Lemma 2.2.1 are fulfilled whenever the claim interarrival times are independent and identically exponentially distributed; that case, however, can be characterized by a much stronger property of the claim number process involving its increments, as will be seen in the following section.

Problem

2.2.A **Discrete Time Model:** Let $\vartheta \in (0, 1)$. Then the following are equivalent:
   (a) $P_{T_n} = \mathbf{Geo}(n, \vartheta)$ for all $n \in \mathbf{N}$.
   (b) $P_{N_l} = \mathbf{B}(l, \vartheta)$ for all $l \in \mathbf{N}$.
   In this case, $E[T_n] = n/\vartheta$ holds for all $n \in \mathbf{N}$ and $E[N_l] = l\vartheta$ holds for all $l \in \mathbf{N}$; moreover, for each $l \in \mathbf{N}$, the pair $(N_l - N_{l-1}, N_{l-1})$ is independent and satisfies $P_{N_l - N_{l-1}} = \mathbf{B}(\vartheta)$.

2.3   A Characterization of the Poisson Process

The claim number process $\{N_t\}_{t \in \mathbf{R}_+}$ has
- *independent increments* if, for all $m \in \mathbf{N}$ and $t_0, t_1, \ldots, t_m \in \mathbf{R}_+$ such that $0 = t_0 < t_1 < \ldots < t_m$, the family of increments $\{N_{t_j} - N_{t_{j-1}}\}_{j \in \{1, \ldots, m\}}$ is independent, it has
- *stationary increments* if, for all $m \in \mathbf{N}$ and $t_0, t_1, \ldots, t_m, h \in \mathbf{R}_+$ such that $0 = t_0 < t_1 < \ldots < t_m$, the family of increments $\{N_{t_j + h} - N_{t_{j-1}+h}\}_{j \in \{1, \ldots, m\}}$ has the same distribution as $\{N_{t_j} - N_{t_{j-1}}\}_{j \in \{1, \ldots, m\}}$, and it is
- a *(homogeneous) Poisson process* with parameter $\alpha \in (0, \infty)$ if it has stationary independent increments such that $P_{N_t} = \mathbf{P}(\alpha t)$ holds for all $t \in (0, \infty)$.

It is immediate from the definitions that a claim number process having independent increments has stationary increments if and only if the identity $P_{N_{t+h} - N_t} = P_{N_h}$ holds for all $t, h \in \mathbf{R}_+$.

The following result exhibits a property of the Poisson process which is not captured by Lemma 2.2.1:

2.3.1 Lemma (Multinomial Criterion). *Let $\alpha \in (0, \infty)$. Then the following are equivalent:*

(a) *The claim number process $\{N_t\}_{t \in \mathbf{R}_+}$ satisfies*

$$P_{N_t} = \mathbf{P}(\alpha t)$$

for all $t \in (0, \infty)$ as well as

$$P\left[\bigcap_{j=1}^{m}\{N_{t_j} - N_{t_{j-1}} = k_j\} \,\middle|\, \{N_{t_m} = n\}\right] = \frac{n!}{\prod_{j=1}^{m} k_j!} \prod_{j=1}^{m}\left(\frac{t_j - t_{j-1}}{t_m}\right)^{k_j}$$

for all $m \in \mathbf{N}$ and $t_0, t_1, \ldots, t_m \in \mathbf{R}_+$ such that $0 = t_0 < t_1 < \ldots < t_m$ and for all $n \in \mathbf{N}_0$ and $k_1, \ldots, k_m \in \mathbf{N}_0$ such that $\sum_{j=1}^{m} k_j = n$.

(b) *The claim number process $\{N_t\}_{t \in \mathbf{R}_+}$ is a Poisson process with parameter α.*

Proof. The result is obtained by straightforward calculation:
- Assume first that (a) holds. Then we have

$$P\left[\bigcap_{j=1}^{m}\{N_{t_j} - N_{t_{j-1}} = k_j\}\right]$$

$$= P\left[\bigcap_{j=1}^{m}\{N_{t_j} - N_{t_{j-1}} = k_j\} \,\middle|\, \{N_{t_m} = n\}\right] \cdot P[\{N_{t_m} = n\}]$$

$$= \frac{n!}{\prod_{j=1}^{m} k_j!} \cdot \prod_{j=1}^{m}\left(\frac{t_j - t_{j-1}}{t_m}\right)^{k_j} \cdot e^{-\alpha t_m}\frac{(\alpha t_m)^n}{n!}$$

$$= \frac{n!}{\prod_{j=1}^{m} k_j!} \cdot \prod_{j=1}^{m} \left(\frac{t_j - t_{j-1}}{t_m} \right)^{k_j} \cdot \prod_{j=1}^{m} e^{-\alpha(t_j - t_{j-1})} \alpha^{k_j} \cdot \frac{t_m^n}{n!}$$

$$= \prod_{j=1}^{m} e^{-\alpha(t_j - t_{j-1})} \frac{(\alpha(t_j - t_{j-1}))^{k_j}}{k_j!} .$$

Therefore, (a) implies (b).

● Assume now that (b) holds. Then we have

$$P_{N_t} = \mathbf{P}(\alpha t)$$

as well as

$$P\left[\bigcap_{j=1}^{m} \{N_{t_j} - N_{t_{j-1}} = k_j\} \middle| \{N_{t_m} = n\} \right] = \frac{P\left[\bigcap_{j=1}^{m} \{N_{t_j} - N_{t_{j-1}} = k_j\} \right]}{P[\{N_{t_m} = n\}]}$$

$$= \frac{\prod_{j=1}^{m} P[\{N_{t_j} - N_{t_{j-1}} = k_j\}]}{P[\{N_{t_m} = n\}]}$$

$$= \frac{\prod_{j=1}^{m} e^{-\alpha(t_j - t_{j-1})} \frac{(\alpha(t_j - t_{j-1}))^{k_j}}{k_j!}}{e^{-\alpha t_m} \frac{(\alpha t_m)^n}{n!}}$$

$$= \frac{n!}{\prod_{j=1}^{m} k_j!} \prod_{j=1}^{m} \left(\frac{t_j - t_{j-1}}{t_m} \right)^{k_j} .$$

Therefore, (b) implies (a).   □

Comparing the previous result with Lemmas 2.2.1 and 1.2.2 raises the question whether the Poisson process can also be characterized in terms of the claim arrival process or in terms of the claim interarrival process. An affirmative answer to this question will be given in Theorem 2.3.4 below.

While the previous result characterizes the Poisson process with parameter α in the class of all claim number processes satisfying $P_{N_t} = \mathbf{P}(\alpha t)$ for all $t \in (0, \infty)$, we shall see that there is also a strikingly simple characterization of the Poisson process in the class of all claim number processes having independent increments; see again Theorem 2.3.4 below.

Theorem 2.3.4 contains two further characterizations of the Poisson process: one in terms of the claim measure, and one in terms of martingales, which are defined as follows:

Let $\mathbf{I}$ be any subset of $\mathbf{R}_+$ and consider a family $\{Z_i\}_{i\in\mathbf{I}}$ of random variables having finite expectations and an increasing family $\{\mathcal{F}_i\}_{i\in\mathbf{I}}$ of sub-σ-algebras of $\mathcal{F}$ such that each Z_i is $\mathcal{F}_i$-measurable. The family $\{\mathcal{F}_i\}_{i\in\mathbf{I}}$ is said to be a *filtration*, and it is said to be the *canonical filtration* for $\{Z_i\}_{i\in\mathbf{I}}$ if it satisfies $\mathcal{F}_i = \sigma(\{Z_h\}_{h\in\mathbf{I}\cap(-\infty,i]})$ for all $i\in\mathbf{I}$. The family $\{Z_i\}_{i\in\mathbf{I}}$ is a
- *submartingale* for $\{\mathcal{F}_i\}_{i\in\mathbf{I}}$ if it satisfies

$$\int_A Z_i\, dP \;\leq\; \int_A Z_j\, dP$$

for all $i,j \in \mathbf{I}$ such that $i < j$ and for all $A \in \mathcal{F}_i$, it is a
- *supermartingale* for $\{\mathcal{F}_i\}_{i\in\mathbf{I}}$ if it satisfies

$$\int_A Z_i\, dP \;\geq\; \int_A Z_j\, dP$$

for all $i,j \in \mathbf{I}$ such that $i < j$ and for all $A \in \mathcal{F}_i$, and it is a
- *martingale* for $\{\mathcal{F}_i\}_{i\in\mathbf{I}}$ if it satisfies

$$\int_A Z_i\, dP \;=\; \int_A Z_j\, dP$$

for all $i,j \in \mathbf{I}$ such that $i < j$ and for all $A \in \mathcal{F}_i$.

Thus, a martingale is at the same time a submartingale and a supermartingale, and all random variables forming a martingale have the same expectation. Reference to the canonical filtration for $\{Z_i\}_{i\in\mathbf{I}}$ is usually omitted.

Let us now return to the claim number process $\{N_t\}_{t\in\mathbf{R}_+}$. For the remainder of this section, let $\{\mathcal{F}_t\}_{t\in\mathbf{R}_+}$ denote the canonical filtration for the claim number process.

The following result connects claim number processes having independent increments and finite expectations with a martingale property:

2.3.2 Theorem.   *Assume that the claim number process $\{N_t\}_{t\in\mathbf{R}_+}$ has independent increments and finite expectations. Then the centered claim number process $\{N_t - E[N_t]\}_{t\in\mathbf{R}_+}$ is a martingale.*

Proof.   Since constants are measurable with respect to any σ-algebra, the natural filtration for the claim number process coincides with the natural filtration for the centered claim number process. Consider $s,t \in \mathbf{R}_+$ such that $s < t$.
(1) *The σ-algebras $\mathcal{F}_s$ and $\sigma(N_t - N_s)$ are independent:*
For $m\in\mathbf{N}$ and $s_0, s_1, \ldots, s_m, s_{m+1} \in \mathbf{R}_+$ such that $0 = s_0 < s_1 < \ldots < s_m = s < t = s_{m+1}$, define

$$
\begin{aligned}
\mathcal{G}_{s_1,\ldots,s_m} \;&:=\; \sigma\big(\{N_{s_j}\}_{j\in\{1,\ldots,m\}}\big) \\
&=\; \sigma\big(\{N_{s_j} - N_{s_{j-1}}\}_{j\in\{1,\ldots,m\}}\big)\,.
\end{aligned}
$$

By assumption, the increments $\{N_{s_j} - N_{s_{j-1}}\}_{j \in \{1,...,m+1\}}$ are independent, and this implies that the σ–algebras $\mathcal{G}_{s_1,...,s_m}$ and $\sigma(N_t - N_s)$ are independent. The system of all such σ–algebras $\mathcal{G}_{s_1,...,s_m}$ is directed upwards by inclusion. Let $\mathcal{E}_s$ denote the union of these σ–algebras. Then $\mathcal{E}_s$ and $\sigma(N_t - N_s)$ are independent. Moreover, $\mathcal{E}_s$ is an algebra, and it follows that $\sigma(\mathcal{E}_s)$ and $\sigma(N_t - N_s)$ are independent. Since $\mathcal{F}_s = \sigma(\mathcal{E}_s)$, this means that the σ–algebras $\mathcal{F}_s$ and $\sigma(N_t - N_s)$ are independent.

(2) Consider now $A \in \mathcal{F}_s$. Because of (1), we have

$$\int_A \left((N_t - E[N_t]) - (N_s - E[N_s]) \right) dP = \int_\Omega \chi_A \left((N_t - N_s) - E[N_t - N_s] \right) dP$$

$$= \int_\Omega \chi_A \, dP \cdot \int_\Omega \left((N_t - N_s) - E[N_t - N_s] \right) dP$$

$$= 0 \,,$$

and hence

$$\int_A (N_t - E[N_t]) \, dP = \int_A (N_s - E[N_s]) \, dP \,.$$

(3) It now follows from (2) that $\{N_t - E[N_t]\}_{t \in \mathbf{R}_+}$ is a martingale.   □

As an immediate consequence of the previous result, we have the following:

2.3.3 Corollary.   *Assume that the claim number process $\{N_t\}_{t \in \mathbf{R}_+}$ is a Poisson process with parameter α. Then the centered claim number process $\{N_t - \alpha t\}_{t \in \mathbf{R}_+}$ is a martingale.*

We shall see that the previous result can be considerably improved.

We now turn to the main result of this section which provides characterizations of the Poisson process in terms of
- the claim interarrival process,
- the increments and expectations of the claim number process,
- the martingale property of a related process, and
- the claim measure.

With regard to the claim measure, we need the following definitions: Define

$$\mathcal{E} := \{A \times (s, t] \mid s, t \in \mathbf{R}_+, s \leq t, A \in \mathcal{F}_s\}$$

and let

$$\mathcal{H} := \sigma(\mathcal{E})$$

denote the σ–algebra generated by $\mathcal{E}$ in $\mathcal{F} \otimes \mathcal{B}((0, \infty))$.

2.3.4 Theorem. *Let $\alpha \in (0, \infty)$. Then the following are equivalent*:
(a) *The sequence of claim interarrival times $\{W_n\}_{n\in\mathbf{N}}$ is independent and satisfies $P_{W_n} = \mathbf{Exp}(\alpha)$ for all $n \in \mathbf{N}$.*
(b) *The claim number process $\{N_t\}_{t\in\mathbf{R}_+}$ is a Poisson process with parameter α.*
(c) *The claim number process $\{N_t\}_{t\in\mathbf{R}_+}$ has independent increments and satisfies $E[N_t] = \alpha t$ for all $t \in \mathbf{R}_+$.*
(d) *The process $\{N_t - \alpha t\}_{t\in\mathbf{R}_+}$ is a martingale.*
(e) *The claim measure μ satisfies $\mu|_{\mathcal{H}} = (\alpha P \otimes \boldsymbol{\lambda})|_{\mathcal{H}}$.*

Proof. We prove the assertion according to the following scheme:

$$(a) \implies (b) \implies (c) \implies (d) \implies (e) \implies (a)$$

• Assume first that (a) holds. The basic idea of this part of the proof is to show that the self–similarity of the survival function of the exponential distribution on the interval $(0, \infty)$ implies self similarity of the claim arrival process in the sense that, for any $s \in \mathbf{R}_+$, the claim arrival process describing the occurrence of claims in the interval (s, ∞) has the same properties as the claim arrival process describing the occurrence of claims in the interval $(0, \infty)$ and is independent of N_s.

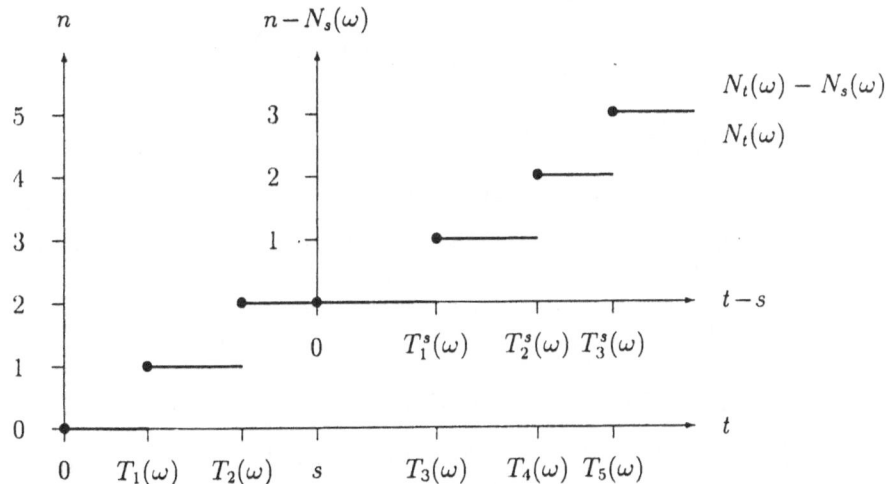

Claim Arrival Process and Claim Number Process

(1) By assumption, the sequence $\{W_n\}_{n\in\mathbf{N}}$ is independent and satisfies

$$P_{W_n} = \mathbf{Exp}(\alpha)$$

for all $n \in \mathbf{N}$. By Lemma 1.2.2, this yields $P_{T_n} = \mathbf{Ga}(\alpha, n)$ for all $n \in \mathbf{N}$, and it now follows from Lemma 2.2.1 that

$$P_{N_t} = \mathbf{P}(\alpha t)$$

holds for all $t \in (0, \infty)$.

(2) Because of (1), we have

$$P[\{N_t = \infty\}] \;=\; 0$$

for all $t \in \mathbf{R}_+$, and it now follows from Lemma 2.1.4 that the probability of explosion is equal to zero. Thus, without loss of generality, we may and do assume that $N_t(\omega) < \infty$ holds for all $t \in \mathbf{R}_+$ and all $\omega \in \Omega$, and this yields

$$\Omega \;=\; \sum_{k=0}^{\infty}\{N_t = k\}$$

for all $t \in \mathbf{R}_+$.

(3) For $s \in \mathbf{R}_+$, define

$$T_0^s \;:=\; 0$$

and, for all $n \in \mathbf{N}$,

$$
\begin{aligned}
T_n^s \;&:=\; \sum_{k=0}^{\infty}\Big(\chi_{\{N_s=k\}} \cdot (T_{k+n}-s)\Big)\\
&=\; \sum_{k=0}^{\infty}\Big(\chi_{\{T_k \leq s < T_{k+1}\}} \cdot (T_{k+n}-s)\Big).
\end{aligned}
$$

Then the sequence $\{T_n^s\}_{n\in\mathbf{N}_0}$ satisfies $T_0^s = 0$ and

$$T_{n-1}^s \;<\; T_n^s$$

for all $n \in \mathbf{N}$. Therefore, $\{T_n^s\}_{n\in\mathbf{N}_0}$ is a claim arrival process. Let $\{W_n^s\}_{n\in\mathbf{N}}$ denote the claim interarrival process induced by $\{T_n^s\}_{n\in\mathbf{N}_0}$.

(4) *For each $s \in \mathbf{R}_+$, the finite dimensional distributions of the claim interarrival processes $\{W_n^s\}_{n\in\mathbf{N}}$ and $\{W_n\}_{n\in\mathbf{N}}$ are identical; moreover, N_s and $\{W_n^s\}_{n\in\mathbf{N}}$ are independent:*

Consider first $t \in \mathbf{R}_+$ and $k \in \mathbf{N}_0$. Then we have

$$
\begin{aligned}
\{N_s = k\}\cap\{t < W_1^s\} \;&=\; \{N_s = k\}\cap\{t < T_1^s\}\\
&=\; \{N_s = k\}\cap\{t < T_{k+1}-s\}\\
&=\; \{T_k \leq s < T_{k+1}\}\cap\{s+t < T_{k+1}\}\\
&=\; \{T_k \leq s\}\cap\{s+t < T_{k+1}\}\\
&=\; \{T_k \leq s\}\cap\{s+t < T_k+W_{k+1}\}\\
&=\; \{T_k \leq s\}\cap\{s-T_k+t < W_{k+1}\}.
\end{aligned}
$$

Using the transformation formula for integrals, independence of T_k and W_{k+1}, and Fubini's theorem, we obtain

$$P[\{N_s = k\}\cap\{t < W_1^s\}] \;=\; P[\{T_k \leq s\}\cap\{s-T_k+t < W_{k+1}\}]$$

$$= \int_\Omega \chi_{\{T_k \le s\} \cap \{s - T_k + t < W_{k+1}\}}(\omega)\, dP(\omega)$$

$$= \int_{\mathbf{R}^2} \chi_{(-\infty,s]}(r)\, \chi_{(s-r+t,\infty)}(u)\, dP_{W_{k+1}, T_k}(u, r)$$

$$= \int_{\mathbf{R}^2} \chi_{(-\infty,s]}(r)\, \chi_{(s-r+t,\infty)}(u)\, d(P_{W_{k+1}} \otimes P_{T_k})(u, r)$$

$$= \int_{\mathbf{R}} \chi_{(-\infty,s]}(r) \left(\int_{\mathbf{R}} \chi_{(s-r+t,\infty)}(u)\, dP_{W_{k+1}}(u) \right) dP_{T_k}(r)$$

$$= \int_{(-\infty,s]} \left(\int_\Omega \chi_{\{s-r+t < W_{k+1}\}}(\omega)\, dP(\omega) \right) dP_{T_k}(r)$$

$$= \int_{(-\infty,s]} P[\{s - r + t < W_{k+1}\}]\, dP_{T_k}(r)\,.$$

Using this formula twice together with the fact that the distribution of each W_n is **Exp**(α) and hence memoryless on $\mathbf{R}_+$, we obtain

$$
\begin{aligned}
P[\{N_s = k\} \cap \{t < W_1^s\}] &= \int_{(-\infty,s]} P[\{s - r + t < W_{k+1}\}]\, dP_{T_k}(r) \\
&= \int_{(-\infty,s]} P[\{s - r < W_{k+1}\}]\, P[\{t < W_{k+1}\}]\, dP_{T_k}(r) \\
&= \int_{(-\infty,s]} P[\{s - r < W_{k+1}\}]\, dP_{T_k}(r) \cdot P[\{t < W_{k+1}\}] \\
&= P[\{N_s = k\} \cap \{0 < W_1^s\}] \cdot P[\{t < W_{k+1}\}] \\
&= P[\{N_s = k\}] \cdot P[\{t < W_1\}]\,.
\end{aligned}
$$

Therefore, we have

$$P[\{N_s = k\} \cap \{t < W_1^s\}] = P[\{N_s = k\}] \cdot P[\{t < W_1\}]\,.$$

Consider now $n \in \mathbf{N}$, $t_1, \ldots, t_n \in \mathbf{R}_+$, and $k \in \mathbf{N}_0$. For each $j \in \{2, \ldots, n\}$, we have

$$
\begin{aligned}
\{N_s = k\} \cap \{t_j < W_j^s\} &= \{N_s = k\} \cap \{t_j < T_j^s - T_{j-1}^s\} \\
&= \{N_s = k\} \cap \{t_j < T_{k+j} - T_{k+j-1}\} \\
&= \{N_s = k\} \cap \{t_j < W_{k+j}\}\,.
\end{aligned}
$$

Since the sequence $\{W_n\}_{n \in \mathbf{N}}$ is independent and identically distributed, the previous identities yield

$$P\left[\{N_s = k\} \cap \bigcap_{j=1}^n \{t_j < W_j^s\} \right]$$

$$= P\left[\bigcap_{j=1}^n \big(\{N_s = k\} \cap \{t_j < W_j^s\} \big) \right]$$

$$= P\left[\left(\{N_s = k\}\cap\{t_1 < W_1^s\}\right)\cap\bigcap_{j=2}^{n}\left(\{N_s = k\}\cap\{t_j < W_j^s\}\right)\right]$$

$$= P\left[\left(\{N_s = k\}\cap\{t_1 < W_1^s\}\right)\cap\bigcap_{j=2}^{n}\left(\{N_s = k\}\cap\{t_j < W_{k+j}\}\right)\right]$$

$$= P\left[\left(\{N_s = k\}\cap\{t_1 < W_1^s\}\right)\cap\bigcap_{j=2}^{n}\{t_j < W_{k+j}\}\right]$$

$$= P\left[\left(\{T_k \le s\}\cap\{s-T_k+t_1 < W_{k+1}\}\right)\cap\bigcap_{j=2}^{n}\{t_j < W_{k+j}\}\right]$$

$$= P[\{T_k \le s\}\cap\{s-T_k+t_1 < W_{k+1}\}]\cdot\prod_{j=2}^{n}P[\{t_j < W_{k+j}\}]$$

$$= P[\{N_s = k\}\cap\{t_1 < W_1^s\}]\cdot\prod_{j=2}^{n}P[\{t_j < W_{k+j}\}]$$

$$= P[\{N_s = k\}]\cdot P[\{t_1 < W_1\}]\cdot\prod_{j=2}^{n}P[\{t_j < W_j\}]$$

$$= P[\{N_s = k\}]\cdot\prod_{j=1}^{n}P[\{t_j < W_j\}]$$

$$= P[\{N_s = k\}]\cdot P\left[\bigcap_{j=1}^{n}\{t_j < W_j\}\right]\ .$$

Therefore, we have

$$P\left[\{N_s = k\}\cap\bigcap_{j=1}^{n}\{t_j < W_j^s\}\right] = P[\{N_s = k\}]\cdot P\left[\bigcap_{j=1}^{n}\{t_j < W_j\}\right]\ .$$

Summation over $k \in \mathbf{N}_0$ yields

$$P\left[\bigcap_{j=1}^{n}\{t_j < W_j^s\}\right] = P\left[\bigcap_{j=1}^{n}\{t_j < W_j\}\right]\ .$$

Inserting this identity into the previous one, we obtain

$$P\left[\{N_s = k\}\cap\bigcap_{j=1}^{n}\{t_j < W_j^s\}\right] = P[\{N_s = k\}]\cdot P\left[\bigcap_{j=1}^{n}\{t_j < W_j^s\}\right]\ .$$

The last two identities show that the finite dimensional distributions of the claim interarrival processes $\{W_n^s\}_{n\in\mathbf{N}}$ and $\{W_n\}_{n\in\mathbf{N}}$ are identical, and that N_s and $\{W_n^s\}_{n\in\mathbf{N}}$ are independent. In particular, the sequence $\{W_n^s\}_{n\in\mathbf{N}}$ is independent and satisfies $P_{W_n^s} = \mathbf{Exp}(\alpha)$ for all $n \in \mathbf{N}$.

(5) *The identity $P_{N_{s+h}-N_s} = P_{N_h}$ holds for all $s, h \in \mathbf{R}_+$:*
For all $n \in \mathbf{N}_0$, we have

$$
\begin{aligned}
\{N_{s+h} - N_s = n\} &= \sum_{k=0}^{\infty} \{N_s = k\} \cap \{N_{s+h} = k+n\} \\
&= \sum_{k=0}^{\infty} \{N_s = k\} \cap \{T_{k+n} \le s+h < T_{k+n+1}\} \\
&= \sum_{k=0}^{\infty} \{N_s = k\} \cap \{T_n^s \le h < T_{n+1}^s\} \\
&= \{T_n^s \le h < T_{n+1}^s\} \, .
\end{aligned}
$$

Because of (4), the finite dimensional distributions of the claim interarrival processes $\{W_n^s\}_{n\in\mathbf{N}}$ and $\{W_n\}_{n\in\mathbf{N}}$ are identical, and it follows that the finite dimensional distributions of the claim arrival processes $\{T_n^s\}_{n\in\mathbf{N}_0}$ and $\{T_n\}_{n\in\mathbf{N}_0}$ are identical as well. This yields

$$
\begin{aligned}
P[\{N_{s+h} - N_s = n\}] &= P[\{T_n^s \le h < T_{n+1}^s\}] \\
&= P[\{T_n \le h < T_{n+1}\}] \\
&= P[\{N_h = n\}]
\end{aligned}
$$

for all $n \in \mathbf{N}_0$.

(6) *The claim number process $\{N_t\}_{t\in\mathbf{R}_+}$ has independent increments:*
Consider first $s \in \mathbf{R}_+$. Because of (4), N_s and $\{W_n^s\}_{n\in\mathbf{N}}$ are independent and the finite dimensional distributions of the claim interarrival processes $\{W_n^s\}_{n\in\mathbf{N}}$ and $\{W_n\}_{n\in\mathbf{N}}$ are identical; consequently, N_s and $\{T_n^s\}_{n\in\mathbf{N}_0}$ are independent and, as noted before, the finite dimensional distributions of the claim arrival processes $\{T_n^s\}_{n\in\mathbf{N}_0}$ and $\{T_n\}_{n\in\mathbf{N}_0}$ are identical as well.
Consider next $s \in \mathbf{R}_+$, $m \in \mathbf{N}$, $h_1, \ldots, h_m \in \mathbf{R}_+$, and $k, k_1, \ldots, k_m \in \mathbf{N}_0$. Then we have

$$
\begin{aligned}
&P\left[\{N_s = k\} \cap \bigcap_{j=1}^{m} \{N_{s+h_j} - N_s = k_j\}\right] \\
&= P\left[\{N_s = k\} \cap \bigcap_{j=1}^{m} \{T_{k_j}^s \le h_j < T_{k_j+1}^s\}\right] \\
&= P[\{N_s = k\}] \cdot P\left[\bigcap_{j=1}^{m} \{T_{k_j}^s \le h_j < T_{k_j+1}^s\}\right] \\
&= P[\{N_s = k\}] \cdot P\left[\bigcap_{j=1}^{m} \{T_{k_j} \le h_j < T_{k_j+1}\}\right] \\
&= P[\{N_s = k\}] \cdot P\left[\bigcap_{j=1}^{m} \{N_{h_j} = k_j\}\right] \, .
\end{aligned}
$$

We now claim that, for all $m \in \mathbf{N}$, the identity

$$P\left[\bigcap_{j=1}^{m}\{N_{t_j} - N_{t_{j-1}} = n_j\}\right] = \prod_{j=1}^{m} P[\{N_{t_j} - N_{t_{j-1}} = n_j\}].$$

holds for all $t_0, t_1, \ldots, t_m \in \mathbf{R}_+$ such that $0 = t_0 < t_1 < \ldots < t_m$, and $n_1, \ldots, n_m \in \mathbf{N}_0$. This follows by induction:

The assertion is obvious for $m = 1$.

Assume now that it holds for some $m \in \mathbf{N}$ and consider $t_0, t_1, \ldots, t_m, t_{m+1} \in \mathbf{R}_+$ such that $0 = t_0 < t_1 < \ldots < t_m < t_{m+1}$, and $n_1, \ldots, n_m, n_{m+1} \in \mathbf{N}_0$. For $j \in \{0, 1, \ldots, m\}$, define $h_j := t_{j+1} - t_1$. Then we have $0 = h_0 < h_1 < \ldots < h_m$ and hence, by assumption and because of (5),

$$P\left[\bigcap_{j=1}^{m}\{N_{h_j} - N_{h_{j-1}} = n_{j+1}\}\right] = \prod_{j=1}^{m} P[\{N_{h_j} - N_{h_{j-1}} = n_{j+1}\}]$$

$$= \prod_{j=1}^{m} P[\{N_{h_j - h_{j-1}} = n_{j+1}\}]$$

$$= \prod_{j=1}^{m} P[\{N_{t_{j+1} - t_j} = n_{j+1}\}]$$

$$= \prod_{j=1}^{m} P[\{N_{t_{j+1}} - N_{t_j} = n_{j+1}\}]$$

$$= \prod_{j=2}^{m+1} P[\{N_{t_j} - N_{t_{j-1}} = n_j\}].$$

Using the identity established before with $s := t_1$, this yields

$$P\left[\bigcap_{j=1}^{m+1}\{N_{t_j} - N_{t_{j-1}} = n_j\}\right] = P\left[\bigcap_{j=1}^{m+1}\left\{N_{t_j} = \sum_{i=1}^{j} n_i\right\}\right]$$

$$= P\left[\{N_{t_1} = n_1\} \cap \bigcap_{j=2}^{m+1}\left\{N_{t_j} = \sum_{i=1}^{j} n_i\right\}\right]$$

$$= P\left[\{N_{t_1} = n_1\} \cap \bigcap_{j=2}^{m+1}\left\{N_{t_j} - N_{t_1} = \sum_{i=2}^{j} n_i\right\}\right]$$

$$= P\left[\{N_{t_1} = n_1\} \cap \bigcap_{j=1}^{m}\left\{N_{t_{j+1}} - N_{t_1} = \sum_{i=2}^{j+1} n_i\right\}\right]$$

$$= P\left[\{N_{t_1} = n_1\} \cap \bigcap_{j=1}^{m}\left\{N_{t_1+h_j} - N_{t_1} = \sum_{i=2}^{j+1} n_i\right\}\right]$$

$$= P[\{N_{t_1} = n_1\}] \cdot P\left[\bigcap_{j=1}^{m}\left\{N_{h_j} = \sum_{i=2}^{j+1} n_i\right\}\right]$$

$$= P[\{N_{t_1} = n_1\}] \cdot P\left[\bigcap_{j=1}^{m}\{N_{h_j} - N_{h_{j-1}} = n_{j+1}\}\right]$$

$$= P[\{N_{t_1} = n_1\}] \cdot \prod_{j=2}^{m+1} P[\{N_{t_j} - N_{t_{j-1}} = n_j\}]$$

$$= \prod_{j=1}^{m+1} P[\{N_{t_j} - N_{t_{j-1}} = n_j\}] \,,$$

which is the assertion for $m+1$. This proves our claim, and it follows that the claim number process $\{N_t\}_{t \in \mathbf{R}_+}$ has independent increments.

(7) It now follows from (6), (5), and (1) that the claim number process $\{N_t\}_{t \in \mathbf{R}_+}$ is a Poisson process with parameter α. Therefore, (a) implies (b).

• Assume now that (b) holds. Since $\{N_t\}_{t \in \mathbf{R}_+}$ is a Poisson process with parameter α, it is clear that $\{N_t\}_{t \in \mathbf{R}_+}$ has independent increments and satisfies $E[N_t] = \alpha t$ for all $t \in \mathbf{R}_+$. Therefore, (b) implies (c).

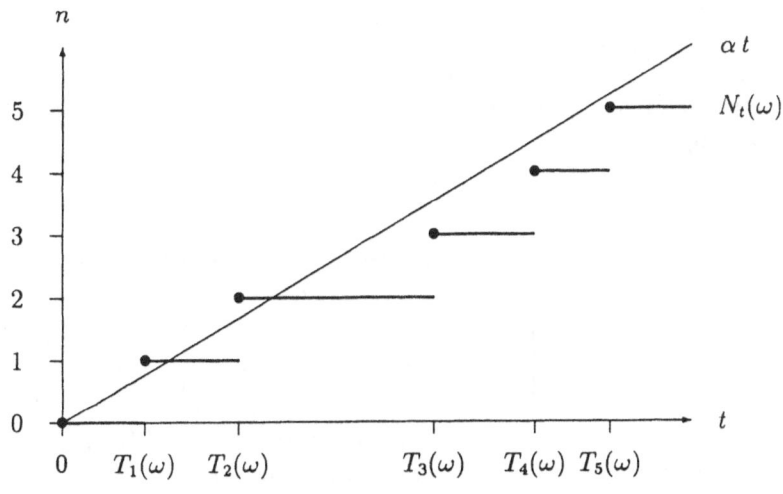

Claim Arrival Process and Claim Number Process

• Assume next that (c) holds. Since $\{N_t\}_{t \in \mathbf{R}_+}$ has independent increments and satisfies $E[N_t] = \alpha t$ for all $t \in \mathbf{R}_+$, it follows from Theorem 2.3.2 that $\{N_t - \alpha t\}_{t \in \mathbf{R}_+}$ is a martingale. Therefore, (c) implies (d).

• Assume now that (d) holds. For all $s, t \in \mathbf{R}_+$ such that $s \leq t$ and for all $A \in \mathcal{F}_s$, Lemma 2.1.6 together with the martingale property of $\{N_r - \alpha r\}_{r \in \mathbf{R}_+}$ yields

$$
\begin{aligned}
\mu[A \times (s, t]] &= \int_A (N_t - N_s) \, dP \\
&= \int_A \alpha(t - s) \, dP \\
&= \alpha(t - s) \cdot P[A] \\
&= \alpha \, \boldsymbol{\lambda}[(s, t]] \cdot P[A] \\
&= (\alpha P \otimes \boldsymbol{\lambda})[A \times (s, t]] \, .
\end{aligned}
$$

Since $A \times (s, t]$ is a typical set of $\mathcal{E}$, this gives

$$
\mu|_{\mathcal{E}} = (\alpha P \otimes \boldsymbol{\lambda})|_{\mathcal{E}} \, .
$$

Since $\Omega \times (0, \infty) = \sum_{n=1}^{\infty} \Omega \times (n-1, n]$, the set $\Omega \times (0, \infty)$ is the union of countably many sets in $\mathcal{E}$ such that $\alpha P \otimes \boldsymbol{\lambda}$ is finite on each of these sets. This means that the measure $\mu|_{\mathcal{E}} = (\alpha P \otimes \boldsymbol{\lambda})|_{\mathcal{E}}$ is σ–finite. Furthermore, since the family $\{\mathcal{F}_s\}_{s \in \mathbf{R}_+}$ is increasing, it is easy to see that $\mathcal{E}$ is stable under intersection. Since $\sigma(\mathcal{E}) = \mathcal{H}$, it now follows from the uniqueness theorem for σ–finite measures that

$$
\mu|_{\mathcal{H}} = (\alpha P \otimes \boldsymbol{\lambda})|_{\mathcal{H}} \, .
$$

Therefore, (d) implies (e)

• Assume finally that (e) holds. In order to determine the finite dimensional distributions of the claim interarrival process $\{W_n\}_{n \in \mathbf{N}}$, we have to study the probability of events having the form

$$
A \cap \{t < W_n\}
$$

for $n \in \mathbf{N}$, $t \in \mathbf{R}_+$, and $A \in \sigma(\{W_k\}_{k \in \{1, \ldots, n-1\}}) = \sigma(\{T_k\}_{k \in \{0, 1, \ldots, n-1\}})$.

(1) For $n \in \mathbf{N}_0$, define

$$
\mathcal{E}_n := \left\{ \bigcap_{k=1}^{n} \{t_k < T_k\} \, \middle| \, t_1, \ldots, t_n \in \mathbf{R}_+ \right\} \, .
$$

Since $\mathcal{E}_n$ is stable under intersection and satisfies $\sigma(\mathcal{E}_n) = \sigma(\{T_k\}_{k \in \{0, 1, \ldots, n\}})$, it is sufficient to study the probability of events having the form

$$
A \cap \{t < W_n\}
$$

for $n \in \mathbf{N}$, $t \in \mathbf{R}_+$, and $A \in \mathcal{E}_{n-1}$.

(2) For $n \in \mathbf{N}$, $t \in \mathbf{R}_+$, and $A \in \mathcal{E}_{n-1}$, define

$$
H_{n,t}(A) := \left\{ (\omega, u) \mid \omega \in A, \, T_{n-1}(\omega) + t < u \leq T_n(\omega) \right\} \, .
$$

Then we have

$$\begin{aligned}
U_n^{-1}(H_{n,t}(A)) &= A \cap \{T_{n-1}+t < T_n\} \\
&= A \cap \{t < W_n\} \,,
\end{aligned}$$

as well as

$$U_k^{-1}(H_{n,t}(A)) = \emptyset$$

for all $k \in \mathbf{N}$ such that $k \neq n$. This gives

$$A \cap \{t < W_n\} = \sum_{k=1}^{\infty} U_k^{-1}(H_{n,t}(A)) \,.$$

Now the problem is to show that $H_{n,t}(A) \in \mathcal{H}$; if this is true, then we can apply the assumption on the claim measure μ in order to compute $P[A \cap \{t < W_n\}]$.

(3) *The relation* $H_{n,t}(A) \in \mathcal{H}$ *holds for all* $n \in \mathbf{N}$, $t \in \mathbf{R}_+$, *and* $A \in \mathcal{E}_{n-1}$:

First, for all $k, m \in \mathbf{N}_0$ such that $k \leq m$ and all $p, q, s, t \in \mathbf{R}_+$ such that $s+t < p < q$, we have

$$\left(\{s < T_k\} \cap \{T_m+t \leq p\}\right) \times (p,q] = \left(\{N_s < k\} \cap \{m \leq N_{p-t}\}\right) \times (p,q] \,,$$

which is a set in $\mathcal{E}$ and hence in $\mathcal{H}$.

Next, for all $k, m \in \mathbf{N}$ such that $k \leq m$ and all $s, t \in \mathbf{R}_+$, define

$$H_{k,m;s,t} := \{(\omega, u) \mid s < T_k(\omega), \ T_m(\omega)+t < u\} \,.$$

Then we have

$$H_{k,m;s,t} = \bigcup_{p,q \in \mathbf{Q}, \, s+t<p<q} \left(\{s < T_k\} \cap \{T_m+t \leq p\}\right) \times (p,q] \,,$$

and hence $H_{k,m;s,t} \in \mathcal{H}$.

Finally, since

$$A = \bigcap_{k=1}^{n-1} \{t_k < T_k\}$$

for suitable $t_1, \ldots, t_{n-1} \in \mathbf{R}_+$, we have

$$\begin{aligned}
H_{n,t}(A) &= H_{n,t}\left(\bigcap_{k=1}^{n-1} \{t_k < T_k\}\right) \\
&= \left\{(\omega, u) \ \middle| \ \omega \in \bigcap_{k=1}^{n-1} \{t_k < T_k\}, \ T_{n-1}(\omega)+t < u \leq T_n(\omega)\right\} \\
&= \bigcap_{k=1}^{n-1} \{(\omega, u) \mid t_k < T_k(\omega), \ T_{n-1}(\omega)+t < u\} \cap \overline{\{(\omega, u) \mid T_n(\omega) < u\}} \\
&= \bigcap_{k=1}^{n-1} H_{k,n-1;t_k,t} \cap \overline{H_{n,n;0,0}} \,,
\end{aligned}$$

and hence $H_{n,t}(A) \in \mathcal{H}$.

(4) Consider $n \in \mathbf{N}$, $t \in \mathbf{R}_+$, and $A \in \mathcal{E}_{n-1}$. Because of (2) and (3), the assumption on the claim measure yields

$$
\begin{aligned}
P[A \cap \{t < W_n\}] &= P\left[\sum_{k=1}^{\infty} U_k^{-1}(H_{n,t}(A))\right] \\
&= \sum_{k=1}^{\infty} P[U_k^{-1}(H_{n,t}(A))] \\
&= \sum_{k=1}^{\infty} P_{U_k}[H_{n,t}(A)] \\
&= \mu[H_{n,t}(A)] \\
&= (\alpha P \otimes \lambda)[H_{n,t}(A)] .
\end{aligned}
$$

Thus, using the fact that the Lebesgue measure is translation invariant and vanishes on singletons, we have

$$
\int_{\mathbf{R}} \chi_{(T_{n-1}(\omega)+t, T_n(\omega)]}(s) \, d\lambda(s) = \int_{\mathbf{R}} \chi_{[t, W_n(\omega))}(s) \, d\lambda(s)
$$

hence

$$
\begin{aligned}
\frac{1}{\alpha} P[A \cap \{t < W_n\}] &= (P \otimes \lambda)[H_{n,t}(A)] \\
&= \int_{\Omega \times \mathbf{R}} \chi_{H_{n,t}(A)}(\omega, s) \, d(P \otimes \lambda)(\omega, s) \\
&= \int_{\Omega \times \mathbf{R}} \chi_A(\omega) \, \chi_{(T_{n-1}(\omega)+t, T_n(\omega)]}(s) \, d(P \otimes \lambda)(\omega, s) \\
&= \int_{\Omega} \chi_A(\omega) \left(\int_{\mathbf{R}} \chi_{(T_{n-1}(\omega)+t, T_n(\omega)]}(s) \, d\lambda(s) \right) dP(\omega) \\
&= \int_{\Omega} \chi_A(\omega) \left(\int_{\mathbf{R}} \chi_{[t, W_n(\omega))}(s) \, d\lambda(s) \right) dP(\omega) \\
&= \int_{\Omega \times \mathbf{R}} \chi_A(\omega) \, \chi_{[t, W_n(\omega))}(s) \, d(P \otimes \lambda)(\omega, s) \\
&= \int_{\Omega \times \mathbf{R}} \chi_{[t, \infty)}(s) \, \chi_{A \cap \{s < W_n\}}(\omega) \, d(P \otimes \lambda)(\omega, s) \\
&= \int_{\mathbf{R}} \chi_{[t, \infty)}(s) \left(\int_{\Omega} \chi_{A \cap \{s < W_n\}} \, dP(\omega) \right) d\lambda(s) \\
&= \int_{[t, \infty)} P[A \cap \{s < W_n\}] \, d\lambda(s) ,
\end{aligned}
$$

and thus

$$
P[A \cap \{t < W_n\}] = \alpha \int_{[t, \infty)} P[A \cap \{s < W_n\}] \, d\lambda(s) .
$$

(5) Consider $n \in \mathbf{N}$ and $A \in \mathcal{E}_{n-1}$. Then the function $g : \mathbf{R} \to \mathbf{R}$, given by

$$g(t) \; := \; \begin{cases} 0 & \text{if} \quad t \in (-\infty, 0) \\ P[A \cap \{t < W_n\}] & \text{if} \quad t \in \mathbf{R}_+ \,, \end{cases}$$

is bounded; moreover, g is monotone decreasing on $\mathbf{R}_+$ and hence almost surely continuous. This implies that g is Riemann integrable and satisfies

$$\int_{[0,t)} g(s) \, d\lambda(s) \;=\; \int_0^t g(s) \, ds$$

for all $t \in \mathbf{R}_+$. Because of (4), the restriction of g to $\mathbf{R}_+$ satisfies

$$\begin{aligned} g(t) \;&=\; P[A \cap \{t < W_n\}] \\ &=\; \alpha \int_{[t,\infty)} P[A \cap \{s < W_n\}] \, d\lambda(s) \\ &=\; \alpha \int_{[t,\infty)} g(s) \, d\lambda(s) \,, \end{aligned}$$

and thus

$$\begin{aligned} g(t) - g(0) \;&=\; -\alpha \int_{[0,t)} g(s) \, d\lambda(s) \\ &=\; -\alpha \int_0^t g(s) \, ds \,. \end{aligned}$$

This implies that the restriction of g to $\mathbf{R}_+$ is differentiable and satisfies the differential equation

$$g'(t) \;=\; -\alpha \, g(t)$$

with initial condition $g(0) = P[A]$.
For all $t \in \mathbf{R}_+$, this yields

$$g(t) \;=\; P[A] \cdot e^{-\alpha t} \,,$$

and thus

$$\begin{aligned} P[A \cap \{t < W_n\}] \;&=\; g(t) \\ &=\; P[A] \cdot e^{-\alpha t} \,. \end{aligned}$$

(6) Consider $n \in \mathbf{N}$. Since $\Omega \in \mathcal{E}_{n-1}$, the previous identity yields

$$P[\{t < W_n\}] \;=\; e^{-\alpha t}$$

for all $t \in \mathbf{R}_+$. Inserting this identity into the previous one, we obtain

$$P[A \cap \{t < W_n\}] \;=\; P[A] \cdot P[\{t < W_n\}]$$

for all $t \in \mathbf{R}_+$ and $A \in \mathcal{E}_{n-1}$. This shows that $\sigma(\{W_1, \ldots, W_{n-1}\})$ and $\sigma(W_n)$ are independent.

(7) Because of (6), it follows by induction that the sequence $\{W_n\}_{n\in\mathbb{N}}$ is independent and satisfies

$$P_{W_n} = \mathbf{Exp}(\alpha)$$

for all $n \in \mathbb{N}$. Therefore, (e) implies (a). $\square$

To conclude this section, let us consider the prediction problem for claim number processes having independent increments and finite second moments:

2.3.5 Theorem (Prediction). *Assume that the claim number process $\{N_t\}_{t\in\mathbb{R}_+}$ has independent increments and finite second moments. Then the inequality*

$$E[(N_t - (N_s + E[N_t - N_s]))^2] \leq E[(N_t - Z)^2]$$

holds for all $s, t \in \mathbb{R}_+$ such that $s \leq t$ and for every random variable Z satisfying $E[Z^2] < \infty$ and $\sigma(Z) \subseteq \mathcal{F}_s$.

Proof. Define $Z_0 := N_s + E[N_t - N_s]$. By assumption, the pair $\{N_t - Z_0, Z_0 - Z\}$ is independent, and this yields

$$\begin{aligned}
E[(N_t - Z_0)(Z_0 - Z)] &= E[N_t - Z_0] \cdot E[Z_0 - Z] \\
&= E[N_t - (N_s + E[N_t - N_s])] \cdot E[Z_0 - Z] \\
&= 0 \,.
\end{aligned}$$

Therefore, we have

$$\begin{aligned}
E[(N_t - Z)^2] &= E[((N_t - Z_0) + (Z_0 - Z))^2] \\
&= E[(N_t - Z_0)^2] + E[(Z_0 - Z)^2] \,.
\end{aligned}$$

The last expression attains its minimum for $Z := Z_0$. $\square$

Thus, for a claim number process having independent increments and finite second moments, the best prediction under expected squared error loss of N_t by a random variable depending only on the history of the claim number process up to time s is given by $N_s + E[N_t - N_s]$.

As an immediate consequence of the previous result, we have the following:

2.3.6 Corollary (Prediction). *Assume that the claim number process $\{N_t\}_{t\in\mathbb{R}_+}$ is a Poisson process with parameter α. Then the inequality*

$$E[(N_t - (N_s + \alpha(t-s)))^2] \leq E[(N_t - Z)^2]$$

holds for all $s, t \in \mathbb{R}_+$ such that $s \leq t$ and for every random variable Z satisfying $E[Z^2] < \infty$ and $\sigma(Z) \subseteq \mathcal{F}_s$.

As in the case of Corollary 2.3.3, the previous result can be considerably improved:

2.3.7 Theorem (Prediction). *Let* $\alpha \in (0, \infty)$. *Then the following are equivalent:*

(a) *The claim number process* $\{N_t\}_{t \in \mathbf{R}_+}$ *has finite second moments and the inequality*

$$E[(N_t - (N_s + \alpha(t-s)))^2] \leq E[(N_t - Z)^2]$$

holds for all $s, t \in \mathbf{R}_+$ *such that* $s \leq t$ *and for every random variable* Z *satisfying* $E[Z^2] < \infty$ *and* $\sigma(Z) \subseteq \mathcal{F}_s$.

(b) *The claim number process* $\{N_t\}_{t \in \mathbf{R}_+}$ *is a Poisson process with parameter* α.

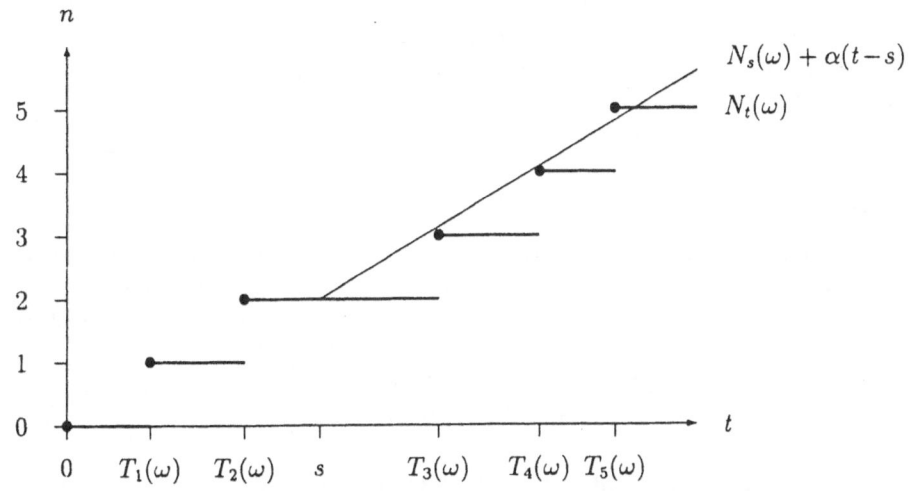

Claim Arrival Process and Claim Number Process

Proof. • Assume that (a) holds and consider $s, t \in \mathbf{R}_+$. For $A \in \mathcal{F}_s$ satisfying $P[A] > 0$, define $Z := N_s + \alpha(t-s) + c\chi_A$. Then we have $\sigma(Z) \subseteq \mathcal{F}_s$, and hence

$$\begin{aligned}
&E[(N_t - (N_s + \alpha(t-s)))^2] \\
&\leq\ E[(N_t - Z)^2] \\
&=\ E[(N_t - (N_s + \alpha(t-s) + c\chi_A))^2] \\
&=\ E[(N_t - (N_s + \alpha(t-s)) - c\chi_A)^2] \\
&=\ E[(N_t - (N_s + \alpha(t-s)))^2] - 2cE[(N_t - (N_s + \alpha(t-s)))\chi_A] + c^2 P[A]\ .
\end{aligned}$$

Letting

$$c\ :=\ \frac{1}{P[A]}\, E[(N_t - (N_s + \alpha(t-s)))\chi_A]\ ,$$

we obtain

$$\begin{aligned}
&E[(N_t - (N_s + \alpha(t-s)))^2] \\
&\leq\ E[(N_t - (N_s + \alpha(t-s)))^2] - \frac{1}{P[A]}\Big(E[(N_t - (N_s + \alpha(t-s)))\chi_A]\Big)^2\ ,
\end{aligned}$$

hence

$$E[(N_t - (N_s + \alpha(t-s))) \chi_A] = 0 \,,$$

and thus

$$\int_A \Big((N_t - \alpha t) - (N_s - \alpha s) \Big) dP = \int_A \Big(N_t - (N_s + \alpha(t-s)) \Big) dP$$
$$= 0 \,.$$

Of course, the previous identity is also valid for $A \in \mathcal{F}_s$ satisfying $P[A] = 0$.
This shows that the process $\{N_t - \alpha t\}_{t \in \mathbf{R}_+}$ is a martingale, and it now follows from
Theorem 2.3.4 that the claim number process $\{N_t\}_{t \in \mathbf{R}_+}$ is a Poisson process with
parameter α. Therefore, (a) implies (b).
• The converse implication is obvious from Corollary 2.3.6.   □

Problems

2.3.A **Discrete Time Model:** Adopt the definitions given in this section to the discrete time model. The claim number process $\{N_l\}_{l \in \mathbf{N}_0}$ is a *binomial process*
or *Bernoulli process* with parameter $\vartheta \in (0,1)$ if it has stationary independent
increments such that $P_{N_l} = \mathbf{B}(l, \vartheta)$ holds for all $l \in \mathbf{N}$.

2.3.B **Discrete Time Model:** Assume that the claim number process $\{N_l\}_{l \in \mathbf{N}_0}$ has independent increments. Then the centered claim number process $\{N_l - E[N_l]\}_{l \in \mathbf{N}_0}$
is a martingale.

2.3.C **Discrete Time Model:** Let $\vartheta \in (0,1)$. Then the following are equivalent:
(a) The claim number process $\{N_l\}_{l \in \mathbf{N}_0}$ satisfies

$$P_{N_l} = \mathbf{B}(l, \vartheta)$$

for all $l \in \mathbf{N}$ as well as

$$P\left[\left. \bigcap_{j=1}^{m} \{N_j - N_{j-1} = k_j\} \, \right| \{N_m = n\} \right] = \binom{m}{n}^{-1}$$

for all $m \in \mathbf{N}$ and for all $n \in \mathbf{N}_0$ and $k_1, \ldots, k_m \in \{0,1\}$ such that $\sum_{j=1}^{m} k_j = n$.
(b) The claim number process $\{N_l\}_{l \in \mathbf{N}_0}$ satisfies

$$P_{N_l} = \mathbf{B}(l, \vartheta)$$

for all $l \in \mathbf{N}$ as well as

$$P\left[\left. \bigcap_{j=1}^{m} \{N_{l_j} - N_{l_{j-1}} = k_j\} \, \right| \{N_{l_m} = n\} \right] = \prod_{j=1}^{m} \binom{l_j - l_{j-1}}{k_j} \cdot \binom{l_m}{n}^{-1}$$

for all $m \in \mathbf{N}$ and $l_0, l_1, \ldots, l_m \in \mathbf{N}_0$ such that $0 = l_0 < l_1 < \ldots < l_m$
and for all $n \in \mathbf{N}_0$ and $k_1, \ldots, k_m \in \mathbf{N}_0$ such that $k_j \leq l_j - l_{j-1}$ for all
$j \in \{1, \ldots, m\}$ and $\sum_{j=1}^{m} k_j = n$.
(c) The claim number process $\{N_l\}_{l \in \mathbf{N}_0}$ is a binomial process with parameter ϑ.

2.3.D **Discrete Time Model:** Let $\vartheta \in (0,1)$. Then the following are equivalent:
    (a) The sequence of claim interarrival times $\{W_n\}_{n\in\mathbf{N}}$ is independent and satisfies $P_{W_n} = \mathbf{Geo}(\vartheta)$ for all $n \in \mathbf{N}$.
    (b) The claim number process $\{N_l\}_{l\in\mathbf{N}_0}$ is a binomial process with parameter ϑ.
    (c) The claim number process $\{N_l\}_{l\in\mathbf{N}_0}$ has independent increments and satisfies $E[N_l] = \vartheta l$ for all $l \in \mathbf{N}_0$.
    (d) The process $\{N_l - \vartheta l\}_{l\in\mathbf{N}_0}$ is a martingale.
    Hint: Prove that (a) $\Longleftrightarrow$ (b) $\Longleftrightarrow$ (c) $\Longleftrightarrow$ (d).

2.3.E **Discrete Time Model:** Assume that the claim number process $\{N_l\}_{l\in\mathbf{N}_0}$ has independent increments. Then the inequality

$$E[(N_m - (N_l + E[N_m - N_l]))^2] \leq E[(N_l - Z)^2]$$

holds for all $l, m \in \mathbf{N}_0$ such that $l \leq m$ and for every random variable Z satisfying $\sigma(Z) \subseteq \mathcal{F}_l$.

2.3.F **Discrete Time Model:** Let $\vartheta \in (0,1)$. Then the following are equivalent:
    (a) The inequality

$$E[(N_m - (N_l + \vartheta(m-l)))^2] \leq E[(N_m - Z)^2]$$

    holds for all $l, m \in \mathbf{N}_0$ such that $l \leq m$ and for every random variable Z satisfying $\sigma(Z) \subseteq \mathcal{F}_l$.
    (b) The claim number process $\{N_l\}_{l\in\mathbf{N}_0}$ is a binomial process with parameter ϑ.

2.3.G **Multiple Life Insurance:** Adopt the definitions given in this section to multiple life insurance. Study stationarity and independence of the increments of the process $\{N_t\}_{t\in\mathbf{R}_+}$ as well as the martingale property of $\{N_t - E[N_t]\}_{t\in\mathbf{R}_+}$.

2.3.H **Single Life Insurance:**
    (a) The process $\{N_t\}_{t\in\mathbf{R}_+}$ does not have stationary increments.
    (b) The process $\{N_t\}_{t\in\mathbf{R}_+}$ has independent increments if and only if the distribution of T is degenerate.
    (c) The process $\{N_t - E[N_t]\}_{t\in\mathbf{R}_+}$ is a martingale if and only if the distribution of T is degenerate.

2.4 Remarks

The definition of the increments of the claim number process suggests to define, for each $\omega \in \Omega$ and all $B \in \mathcal{B}(\mathbf{R})$,

$$N(\omega)(B) \;\; := \;\; \sum_{n=1}^{\infty} \chi_{\{T_n \in B\}}(\omega) \; .$$

Then, for each $\omega \in \Omega$, the map $N(\omega) : \mathcal{B}(\mathbf{R}) \to \mathbf{N}_0 \cup \{\infty\}$ is a measure, which is $\sigma-$finite whenever the probability of explosion is equal to zero. This point of view leads to the theory of *point processes*; see Kerstan, Matthes, and Mecke [1974], Grandell

[1977], Neveu [1977], Matthes, Kerstan, and Mecke [1978], Cox and Isham [1980], Brémaud [1981], Kallenberg [1983], Karr [1991], König and Schmidt [1992], Kingman [1993], and Reiss [1993]; see also Mathar and Pfeifer [1990] for an introduction into the subject.

The implication (a) $\Longrightarrow$ (b) of Theorem 2.3.4 can be used to show that the Poisson process does exist: Indeed, Kolmogorov's existence theorem asserts that, for any sequence $\{Q_n\}_{n\in\mathbf{N}}$ of probability measures $\mathcal{B}(\mathbf{R}) \to [0,1]$, there exists a probability space $(\Omega, \mathcal{F}, P)$ and a sequence $\{W_n\}_{n\in\mathbf{N}}$ of random variables $\Omega \to \mathbf{R}$ such that the sequence $\{W_n\}_{n\in\mathbf{N}}$ is independent and satisfies $P_{W_n} = Q_n$ for all $n \in \mathbf{N}$. Letting $T_n := \sum_{k=1}^{n} W_k$ for all $n \in \mathbf{N}_0$ and $N_t := \sum_{n=1}^{\infty} \chi_{\{T_n \le t\}}$ for all $t \in \mathbf{R}_+$, we obtain a claim arrival process $\{T_n\}_{n\in\mathbf{N}_0}$ and a claim number process $\{N_t\}_{t\in\mathbf{R}_+}$. In particular, if $Q_n = \mathbf{Exp}(\alpha)$ holds for all $n \in \mathbf{N}$, then it follows from Theorem 2.3.4 that $\{N_t\}_{t\in\mathbf{R}_+}$ is a Poisson process with parameter α. The implication (d) $\Longrightarrow$ (b) of Theorem 2.3.4 is due to Watanabe [1964]. The proof of the implications (d) $\Longrightarrow$ (e) and (e) $\Longrightarrow$ (a) of Theorem 2.3.4 follows Letta [1984].

Theorems 2.3.2 and 2.3.4 are typical examples for the presence of martingales in canonical situations in risk theory; see also Chapter 7 below.

In the case where the claim interarrival times are independent and identically (but not necessarily exponentially) distributed, the claim arrival process or, equivalently, the claim number process is said to be a *renewal process*; see e.g. Gut [1988], Alsmeyer [1991], Grandell [1991], and Resnick [1992]. This case will be considered, to a limited extent, in Chapter 7 below.

The case where the claim interarrival times are independent and exponentially (but not necessarily identically) distributed will be studied in Section 3.4 below.

We shall return to the Poisson process at various places in this book: The Poisson process occurs as a very special case in the rather analytical theory of regular claim number processes satisfying the Chapman–Kolmogorov equations, which will be developed in Chapter 3, and it also occurs as a degenerate case in the class of mixed Poisson processes, which will be studied in Chapter 4. Moreover, thinning, decomposition, and superposition of Poisson processes, which are important with regard to reinsurance, will be discussed in Chapters 5 and 6.

Chapter 3

The Claim Number Process as a Markov Process

The characterizations of the Poisson process given in the previous chapter show that the Poisson process is a very particular claim number process. In practical situations, however, the increments of the claim number process may fail to be independent or fail to be stationary or fail to be Poisson distributed, and in each of these cases the Poisson process is not appropriate as a model. The failure of the Poisson process raises the need of studying larger classes of claim number processes.

The present chapter provides a systematic discussion of claim number processes whose transition probabilities satisfy the Chapman–Kolmogorov equations and can be computed from a sequence of intensities. The intensities are functions of time, and special attention will be given to the cases where they are all identical or constant.

We first introduce several properties which a claim number process may possess, which are all related to its transition probabilities, and which are all fulfilled by the Poisson process (Section 3.1). We next give a characterization of regularity of claim number processes satisfying the Chapman–Kolmogorov equations (Section 3.2). Our main results characterize claim number processes which are regular Markov processes with intensities which are all identical (Section 3.3) or all constant (Section 3.4). Combining these results we obtain another characterization of the Poisson process (Section 3.5). We also discuss a claim number process with contagion (Section 3.6).

3.1 The Model

Throughout this chapter, let $\{N_t\}_{t\in\mathbf{R}_+}$ be a claim number process, let $\{T_n\}_{n\in\mathbf{N}_0}$ be the claim arrival process induced by the claim number process, and let $\{W_n\}_{n\in\mathbf{N}}$ be the claim interarrival process induced by the claim arrival process.

In the present section we introduce several properties which a claim number process may possess and which are all fulfilled by the Poisson process. We consider two lines of extending the notion of a Poisson process: The first one is based on the observation that, by definition, every Poisson process has independent increments, and this leads to the more general notion of a Markov claim number process and to the even more general one of a claim number process satisfying the Chapman–Kolmogorov equations. The second one, which has a strongly analytical flavour and is quite different from the first, is the notion of a regular claim number process. Regular claim number processes satisfying the Chapman–Kolmogorov equations will provide the general framework for the discussion of various classes of claim number processes and for another characterization of the Poisson process.

The claim number process $\{N_t\}_{t \in \mathbf{R}_+}$ is a *Markov claim number process*, or a *Markov process* for short, if the identity

$$P\left[\{N_{t_{m+1}} = n_{m+1}\} \,\bigg|\, \bigcap_{j=1}^{m} \{N_{t_j} = n_j\}\right] \;=\; P[\{N_{t_{m+1}} = n_{m+1}\}|\{N_{t_m} = n_m\}]$$

holds for all $m \in \mathbf{N}$, $t_1, \ldots, t_m, t_{m+1} \in (0, \infty)$, and $n_1, \ldots, n_m, n_{m+1} \in \mathbf{N}_0$ such that $t_1 < \ldots < t_m < t_{m+1}$ and $P[\bigcap_{j=1}^{m}\{N_{t_j} = n_j\}] > 0$; the conditions imply that $n_1 \leq \ldots \leq n_m$. Moreover, if the claim number process is a Markov process, then the previous identity remains valid if $t_1 = 0$ or $n_j = \infty$ for some $j \in \{1, \ldots, m\}$.

3.1.1 Theorem. *If the claim number process has independent increments, then it is a Markov process.*

Proof.   Consider $m \in \mathbf{N}$, $t_1, \ldots, t_m, t_{m+1} \in (0, \infty)$, and $n_1, \ldots, n_m, n_{m+1} \in \mathbf{N}_0$ such that $t_1 < \ldots < t_m < t_{m+1}$ and $P[\bigcap_{j=1}^{m}\{N_{t_j} = n_j\}] > 0$. Define $t_0 := 0$ and $n_0 := 0$. Since $P[\{N_0 = 0\}] = 1$, we have

$$
\begin{aligned}
P\left[\{N_{t_{m+1}} = n_{m+1}\} \,\bigg|\, \bigcap_{j=1}^{m}\{N_{t_j} = n_j\}\right]
&= \frac{P\left[\bigcap_{j=1}^{m+1}\{N_{t_j} = n_j\}\right]}{P\left[\bigcap_{j=1}^{m}\{N_{t_j} = n_j\}\right]} \\[2ex]
&= \frac{P\left[\bigcap_{j=1}^{m+1}\{N_{t_j} - N_{t_{j-1}} = n_j - n_{j-1}\}\right]}{P\left[\bigcap_{j=1}^{m}\{N_{t_j} - N_{t_{j-1}} = n_j - n_{j-1}\}\right]} \\[2ex]
&= \frac{\prod_{j=1}^{m+1} P[\{N_{t_j} - N_{t_{j-1}} = n_j - n_{j-1}\}]}{\prod_{j=1}^{m} P[\{N_{t_j} - N_{t_{j-1}} = n_j - n_{j-1}\}]} \\[2ex]
&= P[\{N_{t_{m+1}} - N_{t_m} = n_{m+1} - n_m\}]
\end{aligned}
$$

as well as

$$P[\{N_{t_{m+1}} = n_{m+1}\}|\{N_{t_m} = n_m\}] \;=\; P[\{N_{t_{m+1}} - N_{t_m} = n_{m+1} - n_m\}]\,,$$

and thus

$$P\left[\{N_{t_{m+1}} = n_{m+1}\}\,\Big|\,\bigcap_{j=1}^{m}\{N_{t_j} = n_j\}\right] \;=\; P[\{N_{t_{m+1}} = n_{m+1}\}|\{N_{t_m} = n_m\}]\,.$$

Therefore, $\{N_t\}_{t\in\mathbf{R}_+}$ is a Markov process. $\qquad\qquad\square$

3.1.2 Corollary. *If the claim number process is a Poisson process, then it is a Markov process.*

We shall see later that the claim number process may be a Markov process without being a Poisson process.

In the definition of a Markov claim number process we have already encountered the problem of conditional probabilities with respect to null sets. We now introduce some concepts which will allow us to avoid conditional probabilities with respect to null sets:

A pair $(k,r) \in \mathbf{N}_0 \times \mathbf{R}_+$ is *admissible* if either $(k,r) = (0,0)$ or $(k,r) \in \mathbf{N}_0 \times (0,\infty)$.

Let $\mathcal{A}$ denote the set consisting of all $(k,n,r,t) \in \mathbf{N}_0 \times \mathbf{N}_0 \times \mathbf{R}_+ \times \mathbf{R}_+$ such that (k,r) is admissible, $k \leq n$, and $r \leq t$. A map

$$p : \mathcal{A} \to [0,1]$$

is a *transition rule* for the claim number process $\{N_t\}_{t\in\mathbf{R}_+}$ if it satisfies

$$\sum_{n=k}^{\infty} p(k,n,r,t) \;\leq\; 1$$

for each admissible pair (k,r) and all $t \in [r,\infty)$ as well as

$$p(k,n,r,t) \;=\; P[\{N_t = n\}|\{N_r = k\}]$$

for all $(k,n,r,t) \in \mathcal{A}$ such that $P[\{N_r = k\}] > 0$. It is easy to see that a transition rule always exists but need not be unique. However, all subsequent definitions and results involving transition rules will turn out to be independent of the particular choice of the transition rule.

Comment: The inequality occurring in the definition of a transition rule admits strictly positive probability for a jump to infinity in a finite time interval. For example, for each admissible pair (k, r) and all $t \in [r, \infty)$ such that $P[\{N_r = k\}] > 0$, we have

$$P[\{N_t = \infty\}|\{N_r = k\}] \;\; = \;\; 1 - \sum_{n=k}^{\infty} p(k, n, r, t) \,.$$

Also, since every path of a claim number process is increasing in time, there is no return from infinity. In particular, for all $r \in (0, \infty)$ and $t \in [r, \infty)$ such that $P[\{N_r = \infty\}] > 0$, we have

$$P[\{N_t = \infty\}|\{N_r = \infty\}] \;\; = \;\; 1 \,.$$

These observations show that, as in the definition of Markov claim number processes, infinite claim numbers can be disregarded in the definitions of admissible pairs and transition rules.

For a transition rule $p : \mathcal{A} \to [0, 1]$ and $(k, n, r, t) \in \mathcal{A}$, define

$$p_{k,n}(r, t) \;\; := \;\; p(k, n, r, t) \,.$$

The $p_{k,n}(r, t)$ are called the *transition probabilities* of the claim number process $\{N_t\}_{t \in \mathbf{R}_+}$ with respect to the transition rule p. Obviously, the identity

$$p_{n,n}(t, t) \;\; = \;\; 1$$

holds for each admissible pair (n, t) satisfying $P[\{N_t = n\}] > 0$.

The claim number process $\{N_t\}_{t \in \mathbf{R}_+}$ satisfies the *Chapman–Kolmogorov equations* if there exists a transition rule p such that the identity

$$p_{k,n}(r, t) \;\; = \;\; \sum_{m=k}^{n} p_{k,m}(r, s) \, p_{m,n}(s, t)$$

holds for all $(k, n, r, t) \in \mathcal{A}$ and $s \in [r, t]$ such that $P[\{N_r = k\}] > 0$. The validity of the Chapman–Kolmogorov equations is independent of the particular choice of the transition rule: Indeed, for $m \in \{k, \dots, n\}$ such that $P[\{N_s = m\} \cap \{N_r = k\}] > 0$, we have $P[\{N_s = m\}] > 0$, and thus

$$p_{k,m}(r, s) \, p_{m,n}(s, t) \;\; = \;\; P[\{N_t = n\}|\{N_s = m\}] \cdot P[\{N_s = m\}|\{N_r = k\}] \,;$$

also, for $m \in \{k, \dots, n\}$ such that $P[\{N_s = m\} \cap \{N_r = k\}] = 0$, we have $p_{k,m}(r, s) = P[\{N_s = m\}|\{N_r = k\}] = 0$, and thus

$$p_{k,m}(r, s) \, p_{m,n}(s, t) \;\; = \;\; 0 \,,$$

whatever the value of $p_{m,n}(s, t)$ was defined to be.

3.1.3 Theorem. *If the claim number process is a Markov process, then it satisfies the Chapman–Kolmogorov equations.*

Proof. Consider $(k, n, r, t) \in \mathcal{A}$ and $s \in [r, t]$ such that $P[\{N_r = k\}] > 0$. Then we have

$$
\begin{aligned}
p_{k,n}(r, t) &= P[\{N_t = n\}|\{N_r = k\}] \\
&= \sum_{m=k}^{n} P[\{N_t = n\} \cap \{N_s = m\}|\{N_r = k\}] \\
&= \sum_{m=k}^{n} \Big(P[\{N_t = n\}|\{N_s = m\} \cap \{N_r = k\}] \cdot P[\{N_s = m\}|\{N_r = k\}] \Big) \\
&= \sum_{m=k}^{n} \Big(P[\{N_t = n\}|\{N_s = m\}] \cdot P[\{N_s = m\}|\{N_r = k\}] \Big) \\
&= \sum_{m=k}^{n} p_{k,m}(r, s)\, p_{m,n}(s, t) ,
\end{aligned}
$$

where the second and the third sum are to be taken only over those $m \in \{k, \dots, n\}$ for which $P[\{N_s = m\} \cap \{N_r = k\}] > 0$. $\qquad\square$

3.1.4 Corollary. *If the claim number process has independent increments, then it satisfies the Chapman–Kolmogorov equations.*

3.1.5 Corollary. *If the claim number process is a Poisson process, then it satisfies the Chapman–Kolmogorov equations.*

The claim number process $\{N_t\}_{t \in \mathbf{R}_+}$ is *homogeneous* if there exists a transition rule p such that the identity

$$
p_{n,n+k}(s, s+h) = p_{n,n+k}(t, t+h)
$$

holds for all $n, k \in \mathbf{N}_0$ and $s, t, h \in \mathbf{R}_+$ such that (n, s) and (n, t) are admissible and satisfy $P[\{N_s = n\}] > 0$ and $P[\{N_t = n\}] > 0$. Again, homogeneity is independent of the particular choice of the transition rule.

3.1.6 Theorem. *If the claim number process has stationary independent increments, then it is a homogeneous Markov process.*

Proof. By Theorem 3.1.1, the claim number process is a Markov process.
To prove homogeneity, consider $k \in \mathbf{N}_0$ and $h \in \mathbf{R}_+$ and an admissible pair (n, t) satisfying $P[\{N_t = n\}] > 0$. Then we have

$$
\begin{aligned}
p_{n,n+k}(t, t+h) &= P[\{N_{t+h} = n+k\}|\{N_t = n\}] \\
&= P[\{N_{t+h} - N_t = k\}|\{N_t - N_0 = n\}] \\
&= P[\{N_{t+h} - N_t = k\}] \\
&= P[\{N_h - N_0 = k\}] \\
&= P[\{N_h = k\}] .
\end{aligned}
$$

Therefore, $\{N_t\}_{t \in \mathbf{R}_+}$ is homogeneous. $\qquad\square$

3.1.7 Corollary.   *If the claim number process is a Poisson process, then it is a homogeneous Markov process.*

The relations between the different classes of claim number processes considered so far are presented in the following table:

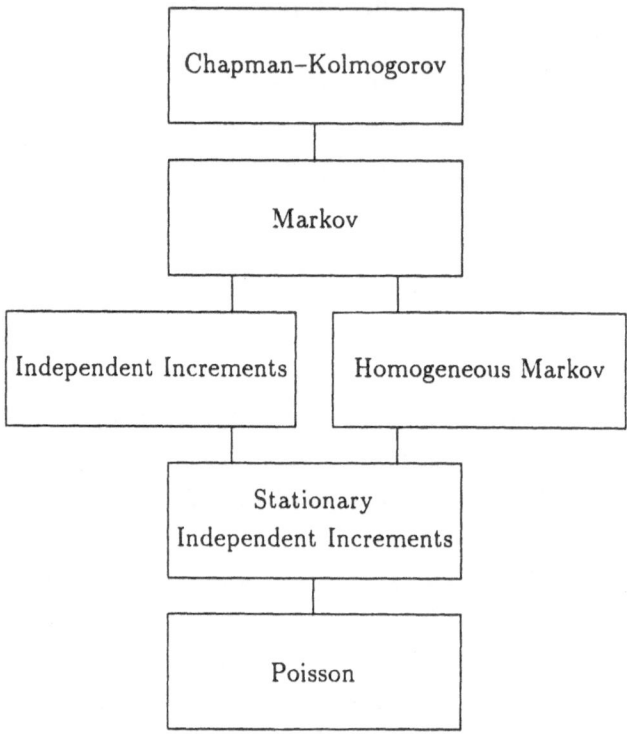

Claim Number Processes

We now turn to another property which a claim number process may possess:

The claim number process $\{N_t\}_{t \in \mathbf{R}_+}$ is *regular* if there exists a transition rule p and a sequence $\{\lambda_n\}_{n \in \mathbf{N}}$ of continuous functions $\mathbf{R}_+ \to (0, \infty)$ such that, for each admissible pair (n, t),

(i)

$$P[\{N_t = n\}] > 0 \,,$$

(ii) the function $\mathbf{R}_+ \to [0, 1] : h \mapsto p_{n,n}(t, t+h)$ is continuous,

(iii)

$$\lim_{h \to 0} \frac{1}{h}\Big(1 - p_{n,n}(t, t+h)\Big) = \lambda_{n+1}(t)$$

$$= \lim_{h \to 0} \frac{1}{h} p_{n,n+1}(t, t+h) \,.$$

In this case, $\{\lambda_n\}_{n\in\mathbb{N}}$ is said to be the *sequence of intensities* of the claim number process. Because of (i), regularity is independent of the particular choice of the transition rule.

Comment:
- Condition (i) means that at any time $t \in (0,\infty)$ every finite number of claims is attained with strictly positive probability.
- Condition (ii) means that, conditionally on the event $\{N_t = n\}$, the probability of no jumps in a finite time interval varies smoothly with the length of the interval.
- Condition (iii) means that, conditionally on the event $\{N_t = n\}$, the tendency for a jump of any height is, in an infinitesimal time interval, equal to the tendency for a jump of height one.

3.1.8 Theorem. *If the claim number process is a Poisson process with parameter α, then it is a homogeneous regular Markov process with intensities $\{\lambda_n\}_{n\in\mathbb{N}}$ satisfying $\lambda_n(t) = \alpha$ for all $n \in \mathbb{N}$ and $t \in \mathbb{R}_+$.*

Proof. By Corollary 3.1.7, the claim number process is a homogeneous Markov process.

To prove the assertion on regularity, consider an admissible pair (n, t).

First, since

$$P[\{N_t = n\}] \;=\; e^{-\alpha t}\,\frac{(\alpha t)^n}{n!}\;,$$

we have $P[\{N_t = n\}] > 0$, which proves (i).

Second, since

$$p_{n,n}(t, t+h) \;=\; e^{-\alpha h}\;,$$

the function $h \mapsto p_{n,n}(t, t+h)$ is continuous, which proves (ii).

Finally, we have

$$\lim_{h\to 0}\frac{1}{h}\Big(1 - p_{n,n}(t, t+h)\Big) \;=\; \lim_{h\to 0}\frac{1}{h}\Big(1 - e^{-\alpha h}\Big)$$
$$=\; \alpha$$

as well as

$$\lim_{h\to 0}\frac{1}{h}\,p_{n,n+1}(t, t+h) \;=\; \lim_{h\to 0}\frac{1}{h}\,e^{-\alpha h}\,\alpha h$$
$$=\; \alpha\;.$$

This proves (iii).

Therefore, $\{N_t\}_{t\in\mathbb{R}_+}$ is regular with intensities $\{\lambda_n\}_{n\in\mathbb{N}}$ satisfying $\lambda_n(t) = \alpha$ for all $n \in \mathbb{N}$ and $t \in \mathbb{R}_+$. $\qquad\square$

The previous result shows that the properties introduced in this section are all fulfilled by the Poisson process.

Problems

3.1.A **Discrete Time Model:** Adopt the definitions given in this section, as far as this is reasonable, to the discrete time model.

3.1.B **Discrete Time Model:** If the claim number process has independent increments, then it is a Markov process.

3.1.C **Discrete Time Model:** If the claim number process is a binomial process, then it is a homogeneous Markov process.

3.1.D **Multiple Life Insurance:** Adopt the definitions given in the section to multiple life insurance. Study the Markov property and regularity of the process $\{N_t\}_{t \in \mathbf{R}_+}$.

3.1.E **Single Life Insurance:** The process $\{N_t\}_{t \in \mathbf{R}_+}$ is a Markov process.

3.1.F **Single Life Insurance:** Assume that $P[\{T > t\}] \in (0, 1)$ holds for all $t \in (0, \infty)$. Then every transition rule p satisfies $p_{0,0}(r, t) = P[\{T > t\}]/P[\{T > r\}]$ as well as $p_{0,1}(r, t) = 1 - p_{0,0}(r, t)$ and $p_{1,1}(r, t) = 1$ for all $r, t \in \mathbf{R}_+$ such that $r \le t$.

3.1.G **Single Life Insurance:** Assume that the distribution of T has a density f with respect to Lebesgue measure and that f is continuous on $\mathbf{R}_+$ and strictly positive on $(0, \infty)$. For all $t \in \mathbf{R}_+$, define

$$\lambda(t) := \frac{f(t)}{P[\{T > t\}]} .$$

(a) The process $\{N_t\}_{t \in \mathbf{R}_+}$ is regular with intensity $\lambda_1 = \lambda$.

(b) There exists a transition rule p such that the differential equations

$$\frac{d}{dt} p_{0,n}(r, t) = \begin{cases} -p_{0,0}(r, t)\, \lambda_1(t) & \text{if } n = 0 \\ p_{0,0}(r, t)\, \lambda_1(t) & \text{if } n = 1 \end{cases}$$

with initial conditions

$$p_{0,n}(r, r) = \begin{cases} 1 & \text{if } n = 0 \\ 0 & \text{if } n = 1 \end{cases}$$

for all $r, t \in \mathbf{R}_+$ such that $r \le t$.

(c) There exists a transition rule p such that the integral equations

$$p_{0,n}(r, t) = \begin{cases} e^{-\int_r^t \lambda_1(s)\, ds} & \text{if } n = 0 \\ \int_r^t p_{0,0}(r, s)\, \lambda_1(s)\, ds & \text{if } n = 1 \end{cases}$$

for all $r, t \in \mathbf{R}_+$ such that $r \le t$.

(d) Interpret the particular form of the differential equations. The function λ is also called the *failure rate* of T.

3.2   A Characterization of Regularity

The following result characterizes regularity of claim number processes satisfying
the Chapman–Kolmogorov equations:

3.2.1 Theorem.   *Assume that the claim number process $\{N_t\}_{t\in\mathbf{R}_+}$ satisfies the
Chapman–Kolmogorov equations and let $\{\lambda_n\}_{n\in\mathbf{N}}$ be a sequence of continuous functions $\mathbf{R}_+ \to (0,\infty)$. Then the following are equivalent:*
(a)   *$\{N_t\}_{t\in\mathbf{R}_+}$ is regular with intensities $\{\lambda_n\}_{n\in\mathbf{N}}$.*
(b)   *There exists a transition rule p such that the differential equations*

$$\frac{d}{dt}\,p_{k,n}(r,t) \;=\; \begin{cases} -\,p_{k,k}(r,t)\,\lambda_{k+1}(t) & \text{if } k=n \\ p_{k,n-1}(r,t)\,\lambda_n(t) - p_{k,n}(r,t)\,\lambda_{n+1}(t) & \text{if } k<n \end{cases}$$

with initial conditions

$$p_{k,n}(r,r) \;=\; \begin{cases} 1 & \text{if } k=n \\ 0 & \text{if } k<n \end{cases}$$

hold for all $(k,n,r,t)\in\mathcal{A}$.
(c)   *There exists a transition rule p such that the integral equations*

$$p_{k,n}(r,t) \;=\; \begin{cases} e^{-\int_r^t \lambda_{k+1}(s)\,ds} & \text{if } k=n \\ \displaystyle\int_r^t p_{k,n-1}(r,s)\,\lambda_n(s)\,p_{n,n}(s,t)\,ds & \text{if } k<n \end{cases}$$

hold for all $(k,n,r,t)\in\mathcal{A}$.

Proof.   We prove the assertion according to the following scheme:

$$\text{(a)} \implies \text{(b)} \implies \text{(c)} \implies \text{(a)}$$

• Assume first that (a) holds and consider a transition rule p and $(k,n,r,t)\in\mathcal{A}$.
(1) By the Chapman–Kolmogorov equations, we have

$$\begin{aligned} p_{k,k}(r,t+h) - p_{k,k}(r,t) &= p_{k,k}(r,t)\,p_{k,k}(t,t+h) - p_{k,k}(r,t) \\ &= -p_{k,k}(r,t)\Big(1 - p_{k,k}(t,t+h)\Big), \end{aligned}$$

and hence

$$\begin{aligned} \lim_{h\to 0}\frac{1}{h}\Big(p_{k,k}(r,t+h) - p_{k,k}(r,t)\Big) &= -p_{k,k}(r,t)\lim_{h\to 0}\frac{1}{h}\Big(1 - p_{k,k}(t,t+h)\Big) \\ &= -p_{k,k}(r,t)\,\lambda_{k+1}(t). \end{aligned}$$

Thus, the right derivative of $t \mapsto p_{k,k}(r,t)$ exists and is continuous, and this implies
that the derivative of $t \mapsto p_{k,k}(r,t)$ exists and satisfies the differential equation

$$\frac{d}{dt}\,p_{k,k}(r,t) \;=\; -p_{k,k}(r,t)\,\lambda_{k+1}(t)$$

with initial condition $p_{k,k}(r,r) = 1$.

In particular, we have

$$p_{k,k}(r,t) = e^{-\int_r^t \lambda_{k+1}(s)\, ds}$$

$$> 0 .$$

(2) Assume now that $k < n$. Then we have

$$p_{k,n}(r,t+h) - p_{k,n}(r,t) = \sum_{m=k}^{n} p_{k,m}(r,t)\, p_{m,n}(t,t+h) - p_{k,n}(r,t)$$

$$= \sum_{m=k}^{n-2} p_{k,m}(r,t)\, p_{m,n}(t,t+h)$$

$$+ p_{k,n-1}(r,t)\, p_{n-1,n}(t,t+h)$$

$$- p_{k,n}(r,t)\Big(1 - p_{n,n}(t,t+h)\Big) .$$

For $m \in \{k,\ldots,n-2\}$, we have $p_{m,n}(t,t+h) \le 1 - p_{m,m}(t,t+h) - p_{m,m+1}(t,t+h)$, hence

$$\lim_{h\to 0} \frac{1}{h}\, p_{m,n}(t,t+h) = 0 ,$$

and this identity together with

$$\lim_{h\to 0} \frac{1}{h}\, p_{n-1,n}(t,t+h) = \lambda_n(t)$$

and

$$\lim_{h\to 0} \frac{1}{h}\Big(1 - p_{n,n}(t,t+h)\Big) = \lambda_{n+1}(t)$$

yields

$$\lim_{h\to 0} \frac{1}{h}\Big(p_{k,n}(r,t+h) - p_{k,n}(r,t)\Big) = \sum_{m=k}^{n-2} p_{k,m}(r,t) \lim_{h\to 0} \frac{1}{h}\, p_{m,n}(t,t+h)$$

$$+ p_{k,n-1}(r,t) \lim_{h\to 0} \frac{1}{h}\, p_{n-1,n}(t,t+h)$$

$$- p_{k,n}(r,t) \lim_{h\to 0} \frac{1}{h}\Big(1 - p_{n,n}(t,t+h)\Big)$$

$$= p_{k,n-1}(r,t)\, \lambda_n(t) - p_{k,n}(r,t)\, \lambda_{n+1}(t)) .$$

Thus, the right derivative of $t \mapsto p_{k,n}(r,t)$ exists, and it follows that the function $t \mapsto p_{k,n}(r,t)$ is right continuous on $[r,\infty)$. Moreover, for $t \in (r,\infty)$ the Chapman-Kolmogorov equations yield, for all $s \in (r,t)$,

$$\Big|p_{k,n}(r,s) - p_{k,n}(r,t)\Big| = \Big|p_{k,n}(r,s) - \sum_{m=k}^{n} p_{k,m}(r,s)\, p_{m,n}(s,t)\Big|$$

$$\leq \sum_{m=k}^{n-1} p_{k,m}(r,s)\, p_{m,n}(s,t) + p_{k,n}(r,s)\Big(1-p_{n,n}(s,t)\Big)$$

$$\leq \sum_{m=k}^{n} \Big(1 - p_{m,m}(s,t)\Big)$$

$$= \sum_{m=k}^{n} \left(1 - \frac{p_{m,m}(r,t)}{p_{m,m}(r,s)}\right),$$

and thus

$$\lim_{s\to t}\big|p_{k,n}(r,s) - p_{k,n}(r,t)\big| = 0,$$

which means that the function $t \mapsto p_{k,n}(r,t)$ is also left continuous on (r,∞) and hence continuous on $[r,\infty)$. But then the right derivative of $t \mapsto p_{k,n}(r,t)$ is continuous, and this implies that the derivative of $t \mapsto p_{k,n}(r,t)$ exists and satisfies the differential equation

$$\frac{d}{dt} p_{k,n}(r,t) = p_{k,n-1}(r,t)\,\lambda_n(t) - p_{k,n}(r,t)\,\lambda_{n+1}(t)$$

with initial condition $p_{k,n}(r,r) = 0$.
(3) Because of (1) and (2), (a) implies (b).
• Assume now that (b) holds and consider a transition rule p satisfying the differential equations and $(k,n,r,t) \in \mathcal{A}$.
(1) We have already noticed in the preceding part of the proof that the differential equation

$$\frac{d}{dt} p_{k,k}(r,t) = -p_{k,k}(r,t)\,\lambda_{k+1}(t)$$

with initial condition $p_{k,k}(r,r) = 1$ has the unique solution

$$p_{k,k}(r,t) = e^{-\int_r^t \lambda_{k+1}(s)\,ds}.$$

(2) Assume now that $k < n$. Then the function $t \mapsto 0$ is the unique solution of the homogeneous differential equation

$$\frac{d}{dt} p_{k,n}(r,t) = -p_{k,n}(r,t)\,\lambda_{n+1}(t)$$

with initial condition $p_{k,n}(r,r) = 0$. This implies that the inhomogeneous differential equation

$$\frac{d}{dt} p_{k,n}(r,t) = p_{k,n-1}(r,t)\,\lambda_n(t) - p_{k,n}(r,t)\,\lambda_{n+1}(t)$$

with initial condition $p_{k,n}(r,r) = 0$ has at most one solution.

Assume that the function $t \mapsto p_{k,n-1}(r,t)$ is already given (which because of (1) is the case for $n = k+1$) and define

$$\hat{p}_{k,n}(r,t) \ := \ \int_r^t p_{k,n-1}(r,s)\,\lambda_n(s)\,p_{n,n}(s,t)\,ds \ .$$

Since

$$\begin{aligned}
\frac{d}{dt}\hat{p}_{k,n}(r,t) \ &= \ \int_r^t p_{k,n-1}(r,s)\,\lambda_n(s)\left(\frac{d}{dt}p_{n,n}(s,t)\right)ds + p_{k,n-1}(r,t)\,\lambda_n(t) \\
&= \ \int_r^t p_{k,n-1}(r,s)\,\lambda_n(s)\left(-p_{n,n}(s,t)\,\lambda_{n+1}(t)\right)ds + p_{k,n-1}(r,t)\,\lambda_n(t) \\
&= \ \int_r^t p_{k,n-1}(r,s)\,\lambda_n(s)\,p_{n,n}(s,t)\,ds \cdot (-\lambda_{n+1}(t)) + p_{k,n-1}(r,t)\,\lambda_n(t) \\
&= \ \hat{p}_{k,n}(r,s)\,(-\lambda_{n+1}(t)) + p_{k,n-1}(r,t)\,\lambda_n(t) \\
&= \ p_{k,n-1}(r,t)\,\lambda_n(t) - \hat{p}_{k,n}(r,s)\,\lambda_{n+1}(t)
\end{aligned}$$

and $\hat{p}_{k,n}(r,r) = 0$, the function $t \mapsto \hat{p}_{k,n}(r,t)$ is the unique solution of the differential equation

$$\frac{d}{dt}p_{k,n}(r,t) \ = \ p_{k,n-1}(r,t)\,\lambda_n(t) - p_{k,n}(r,t)\,\lambda_{n+1}(t) \ ,$$

with initial condition $p_{k,n}(r,r) = 0$, and we have

$$p_{k,n}(r,t) \ := \ \int_r^t p_{k,n-1}(r,s)\,\lambda_n(s)\,p_{n,n}(s,t)\,ds \ .$$

(3) Because of (1) and (2), (b) implies (c).

• Assume finally that (c) holds and consider a transition rule p satisfying the integral equations. For $n \in \mathbf{N}$ and $r,t \in \mathbf{R}_+$ such that $r \le t$, define

$$\Lambda_n(r,t) \ := \ \int_r^t \lambda_n(s)\,ds \ .$$

Then we have

$$\begin{aligned}
p_{n,n}(r,t) \ &= \ e^{-\int_r^t \lambda_{n+1}(s)\,ds} \\
&= \ e^{-\Lambda_{n+1}(r,t)} \\
&> \ 0
\end{aligned}$$

for each admissible pair (n,r) and all $t \in [r,\infty)$.

First, for all $t \in \mathbf{R}_+$, we have

$$\begin{aligned}
P[\{N_t = 0\}] \ &= \ P[\{N_t = 0\}|\{N_0 = 0\}] \\
&= \ p_{0,0}(0,t) \\
&> \ 0 \ .
\end{aligned}$$

Consider now $t \in (0, \infty)$ and assume that $p_{0,n-1}(0,t) = P[\{N_t = n-1\}] > 0$ holds for some $n \in \mathbf{N}$ and all $t \in \mathbf{R}_+$ (which is the case for $n = 1$). Then we have

$$
\begin{aligned}
P[\{N_t = n\}] &= P[\{N_t = n\} | \{N_0 = 0\}] \\
&= p_{0,n}(0,t) \\
&= \int_0^t p_{0,n-1}(0,s) \, \lambda_n(s) \, p_{n,n}(s,t) \, ds \\
&> 0 \, .
\end{aligned}
$$

This yields

$$
P[\{N_t = n\}] > 0
$$

for each admissible pair (n,t), which proves (i).

Second, for each admissible pair (n,t) and for all $h \in \mathbf{R}_+$, we have

$$
p_{n,n}(t, t+h) = e^{-\Lambda_{n+1}(t,t+h)} \, ,
$$

showing that the function $h \mapsto p_{n,n}(t, t+h)$ is continuous, which proves (ii). Finally, for each admissible pair (n,t), we have

$$
\begin{aligned}
\lim_{h \to 0} \frac{1}{h} \Big(1 - p_{n,n}(t, t+h) \Big) &= \lim_{h \to 0} \frac{1}{h} \Big(1 - e^{-\Lambda_{n+1}(t,t+h)} \Big) \\
&= \lambda_{n+1}(t) \, ,
\end{aligned}
$$

and because of

$$
\begin{aligned}
p_{n,n+1}(t, t+h) &= \int_t^{t+h} p_{n,n}(t, u) \, \lambda_{n+1}(u) \, p_{n+1,n+1}(u, t+h) \, du \\
&= \int_t^{t+h} p_{n,n}(t, u) \, \lambda_{n+1}(u) \, e^{-\Lambda_{n+2}(u, t+h)} \, du \\
&= e^{-\Lambda_{n+2}(t, t+h)} \int_t^{t+h} p_{n,n}(t, u) \, \lambda_{n+1}(u) \, e^{\Lambda_{n+2}(t, u)} \, du
\end{aligned}
$$

we also have

$$
\begin{aligned}
& \lim_{h \to 0} \frac{1}{h} p_{n,n+1}(t, t+h) \\
&= \lim_{h \to 0} \frac{1}{h} \left(e^{-\Lambda_{n+2}(t, t+h)} \int_t^{t+h} p_{n,n}(t, u) \, \lambda_{n+1}(u) \, e^{\Lambda_{n+2}(t, u)} \, du \right) \\
&= e^{-\Lambda_{n+2}(t,t)} \cdot p_{n,n}(t,t) \, \lambda_{n+1}(t) \, e^{\Lambda_{n+2}(t,t)} \\
&= \lambda_{n+1}(t) \, .
\end{aligned}
$$

This proves (iii).
Therefore, (c) implies (a). $\qquad\square$

Since regularity is independent of the particular choice of the transition rule, every transition rule for a regular claim number process satisfying the Chapman–Kolmogorov equations fulfills the differential and integral equations of Theorem 3.2.1.

3.3   A Characterization of the Inhomogeneous Poisson Process

In the present section, we study claim number processes which are regular Markov processes with intensities which are all identical.

3.3.1 Lemma.   *Assume that the claim number process $\{N_t\}_{t \in \mathbf{R}_+}$ satisfies the Chapman–Kolmogorov equations and is regular. Then the following are equivalent:*
(a) *The identity*

$$p_{0,k}(t, t+h) \;=\; p_{n,n+k}(t, t+h)$$

holds for all $n, k \in \mathbf{N}_0$ and $t, h \in \mathbf{R}_+$ such that (n, t) is admissible.
(b) *The intensities of $\{N_t\}_{t \in \mathbf{R}_+}$ are all identical.*

Proof. • It is clear that (a) implies (b).
• Assume now that (b) holds.
(1) For each admissible pair (n, t) and all $h \in \mathbf{R}_+$, we have

$$
\begin{aligned}
p_{0,0}(t, t+h) \;&=\; e^{-\int_t^{t+h} \lambda_1(s)\, ds} \\
&=\; e^{-\int_t^{t+h} \lambda_{n+1}(s)\, ds} \\
&=\; p_{n,n}(t, t+h) \,.
\end{aligned}
$$

(2) Assume now that the identity

$$p_{0,k}(t, t+h) \;=\; p_{n,n+k}(t, t+h)$$

holds for some $k \in \mathbf{N}_0$ and for each admissible pair (n, t) and all $h \in \mathbf{R}_+$ (which because of (1) is the case for $k = 0$). Then we have

$$
\begin{aligned}
p_{0,k+1}(t, t+h) \;&=\; \int_t^{t+h} p_{0,k}(t, u)\, \lambda_{k+1}(u)\, p_{k+1,k+1}(u, t+h)\, du \\
&=\; \int_t^{t+h} p_{n,n+k}(t, u)\, \lambda_{n+k+1}(u)\, p_{n+k+1,n+k+1}(u, t+h)\, du \\
&=\; p_{n,n+k+1}(t, t+h) \,.
\end{aligned}
$$

for each admissible pair (n, t) and all $h \in \mathbf{R}_+$.
(3) Because of (1) and (2), (b) implies (a).   □

Let $\lambda : \mathbf{R}_+ \to (0, \infty)$ be a continuous function. The claim number process $\{N_t\}_{t \in \mathbf{R}_+}$ is an *inhomogeneous Poisson process* with *intensity* λ if it has independent increments satisfying

$$P_{N_{t+h}-N_t} \;=\; \mathbf{P}\!\left(\int_t^{t+h} \lambda(s)\, ds \right)$$

for all $t \in \mathbf{R}_+$ and $h \in (0, \infty)$. Thus, the claim number process is an inhomogeneous Poisson process with constant intensity $t \mapsto \alpha$ if and only if it is a Poisson process with parameter α.

3.3.2 Theorem.   *Let $\lambda : \mathbf{R}_+ \to (0, \infty)$ be a continuous function. Then the following are equivalent*:
(a) *The claim number process $\{N_t\}_{t \in \mathbf{R}_+}$ is a regular Markov process with intensities $\{\lambda_n\}_{n \in \mathbf{N}}$ satisfying $\lambda_n(t) = \lambda(t)$ for all $n \in \mathbf{N}$ and $t \in \mathbf{R}_+$.*
(b) *The claim number process $\{N_t\}_{t \in \mathbf{R}_+}$ has independent increments and is regular with intensities $\{\lambda_n\}_{n \in \mathbf{N}}$ satisfying $\lambda_n(t) = \lambda(t)$ for all $n \in \mathbf{N}$ and $t \in \mathbf{R}_+$.*
(c) *The claim number process $\{N_t\}_{t \in \mathbf{R}_+}$ is an inhomogeneous Poisson process with intensity λ.*

Proof.   Each of the conditions implies that the claim number process satisfies the Chapman–Kolmogorov equations. Therefore, Theorem 3.2.1 applies.
For all $r, t \in \mathbf{R}_+$ such that $r \leq t$, define

$$\Lambda(r, t) \; := \; \int_r^t \lambda(s) \, ds \;.$$

We prove the assertion according to the following scheme:

$$(a) \implies (c) \implies (b) \implies (a)$$

• Assume first that (a) holds.
(1) For each admissible pair (n, r) and all $t \in [r, \infty)$, we have

$$
\begin{aligned}
p_{n,n}(r, t) \; &= \; e^{-\int_r^t \lambda_{n+1}(s) \, ds} \\
&= \; e^{-\int_r^t \lambda(s) \, ds} \\
&= \; e^{-\Lambda(r,t)} \;.
\end{aligned}
$$

(2) Assume now that the identity

$$p_{n,n+k}(r, t) \; = \; e^{-\Lambda(r,t)} \frac{(\Lambda(r, t))^k}{k!}$$

holds for some $k \in \mathbf{N}_0$ and for each admissible pair (n, r) and all $t \in [r, \infty)$ (which because of (1) is the case for $k = 0$). Then we have

$$
\begin{aligned}
p_{n,n+k+1}(r, t) \; &= \; \int_r^t p_{n,n+k}(r, s) \, \lambda_{n+k+1}(s) \, p_{n+k+1,n+k+1}(s, t) \, ds \\
&= \; \int_r^t e^{-\Lambda(r,s)} \frac{(\Lambda(r, s))^k}{k!} \lambda(s) \, e^{-\Lambda(s,t)} \, ds \\
&= \; e^{-\Lambda(r,t)} \int_r^t \frac{(\Lambda(r, s))^k}{k!} \lambda(s) \, ds \\
&= \; e^{-\Lambda(r,t)} \frac{(\Lambda(r, t))^{k+1}}{(k+1)!}
\end{aligned}
$$

for each admissible pair (n, r) and all $t \in [r, \infty)$.

(3) Because of (1) and (2), the identity

$$p_{n,n+k}(r,t) = e^{-\Lambda(r,t)} \frac{(\Lambda(r,t))^k}{k!}$$

holds for all $n, k \in \mathbf{N}_0$ and $r, l \in \mathbf{R}_+$ such that (n, r) is admissible and $r \leq t$.

(4) Because of (3), we have

$$\begin{aligned} P[\{N_t = n\}] &= p_{0,n}(0, t) \\ &= e^{-\Lambda(0,t)} \frac{(\Lambda(0,t))^n}{n!} \end{aligned}$$

for all $t \in \mathbf{R}_+$ and $n \in \mathbf{N}_0$ such that (n, t) is admissible, and thus

$$P_{N_t} = \mathbf{P}(\Lambda(0, t)) .$$

for all $t \in (0, \infty)$.

(5) *The identity*

$$P_{N_t - N_r} = \mathbf{P}(\Lambda(r, t))$$

holds for all $r, t \in \mathbf{R}_+$ such that $r < t$.

In the case $r = 0$, the assertion follows from (4).

In the case $r > 0$, it follows from (4) that the probability of explosion is equal to zero and that $P[\{N_r = n\}] > 0$ holds for all $n \in \mathbf{N}_0$. Because of (3), we obtain

$$\begin{aligned} P[\{N_t - N_r = k\}] &= \sum_{n=0}^{\infty} P[\{N_t - N_r = k\} \cap \{N_r = n\}] \\ &= \sum_{n=0}^{\infty} \Big(P[\{N_t = n + k\}|\{N_r = n\}] \cdot P[\{N_r = n\}] \Big) \\ &= \sum_{n=0}^{\infty} p_{n,n+k}(r, t)\, p_{0,n}(0, r) \\ &= \sum_{n=0}^{\infty} e^{-\Lambda(r,t)} \frac{(\Lambda(r,t))^k}{k!}\, e^{-\Lambda(0,r)} \frac{(\Lambda(0,r))^n}{n!} \\ &= e^{-\Lambda(r,t)} \frac{(\Lambda(r,t))^k}{k!} \cdot \sum_{n=0}^{\infty} e^{-\Lambda(0,r)} \frac{(\Lambda(0,r))^n}{n!} \\ &= e^{-\Lambda(r,t)} \frac{(\Lambda(r,t))^k}{k!} \end{aligned}$$

for all $k \in \mathbf{N}_0$, and thus

$$P_{N_t - N_r} = \mathbf{P}(\Lambda(r, t)) .$$

(6) *The claim number process* $\{N_t\}_{t\in\mathbf{R}_+}$ *has independent increments:*
Consider $m \in \mathbf{N}$, $t_0, t_1, \ldots, t_m \in \mathbf{R}_+$ such that $0 = t_0 < t_1 < \ldots < t_m$, and $k_1, \ldots, k_m \in \mathbf{N}_0$. For $j \in \{0, 1, \ldots, m\}$, define

$$n_j := \sum_{i=1}^{j} k_i \,.$$

If $P[\bigcap_{j=1}^{m-1}\{N_{t_j} - N_{t_{j-1}} = k_j\}] = P[\bigcap_{j=1}^{m-1}\{N_{t_j} = n_j\}] > 0$, then the Markov property together with (3) and (5) yields

$$
P\left[\{N_{t_m} = n_m\} \,\middle|\, \bigcap_{j=1}^{m-1}\{N_{t_j} = n_j\}\right]
\;=\; P[\{N_{t_m} = n_m\}|\{N_{t_{m-1}} = n_{m-1}\}]
$$

$$
= \; p_{n_{m-1}, n_m}(t_{m-1}, t_m)
$$

$$
= \; e^{-\Lambda(t_{m-1}, t_m)} \frac{(\Lambda(t_{m-1}, t_m))^{n_m - n_{m-1}}}{(n_m - n_{m-1})!}
$$

$$
= \; P[\{N_{t_m} - N_{t_{m-1}} = n_m - n_{m-1}\}]
$$

$$
= \; P[\{N_{t_m} - N_{t_{m-1}} = k_m\}]
$$

and hence

$$
P\left[\bigcap_{j=1}^{m}\{N_{t_j} - N_{t_{j-1}} = k_j\}\right]
\;=\; P\left[\bigcap_{j=1}^{m}\{N_{t_j} = n_j\}\right]
$$

$$
= \; P\left[\{N_{t_m} = n_m\} \,\middle|\, \bigcap_{j=1}^{m-1}\{N_{t_j} = n_j\}\right] \cdot P\left[\bigcap_{j=1}^{m-1}\{N_{t_j} = n_j\}\right]
$$

$$
= \; P[\{N_{t_m} - N_{t_{m-1}} = k_m\}] \cdot P\left[\bigcap_{j=1}^{m-1}\{N_{t_j} = n_j\}\right]
$$

$$
= \; P[\{N_{t_m} - N_{t_{m-1}} = k_m\}] \cdot P\left[\bigcap_{j=1}^{m-1}\{N_{t_j} - N_{t_{j-1}} = k_j\}\right] \,.
$$

Obviously, the identity

$$
P\left[\bigcap_{j=1}^{m}\{N_{t_j} - N_{t_{j-1}} = k_j\}\right]
\;=\; P[\{N_{t_m} - N_{t_{m-1}} = k_m\}] \cdot P\left[\bigcap_{j=1}^{m-1}\{N_{t_j} - N_{t_{j-1}} = k_j\}\right]
$$

is also valid if $P[\bigcap_{j=1}^{m-1}\{N_{t_j} - N_{t_{j-1}} = k_j\}] = 0$.
It now follows by induction that $\{N_t\}_{t\in\mathbf{R}_+}$ has independent increments.
(7) Because of (5) and (6), $\{N_t\}_{t\in\mathbf{R}_+}$ is an inhomogeneous Poisson process with intensity λ. Therefore, (a) implies (c).

• Assume now that (c) holds. Of course, $\{N_t\}_{t\in\mathbf{R}_+}$ has independent increments. Furthermore, for each admissible pair (n,r) and all $k \in \mathbf{N}_0$ and $t \in [r,\infty)$, we have

$$
\begin{aligned}
p_{n,n+k}(r,t) &= P[\{N_t = n+k\}|\{N_r = n\}] \\
&= P[\{N_t - N_r = k\}|\{N_r - N_0 = n\}] \\
&= P[\{N_t - N_r = k\}] \\
&= e^{-\Lambda(r,t)} \frac{(\Lambda(r,t))^k}{k!} \ .
\end{aligned}
$$

For $k = 0$, this yields

$$
\begin{aligned}
p_{n,n}(r,t) &= e^{-\Lambda(r,t)} \\
&= e^{-\int_r^t \lambda(s)\, ds} \ ,
\end{aligned}
$$

and for $k \in \mathbf{N}$ we obtain

$$
\begin{aligned}
p_{n,n+k}(r,t) &= e^{-\Lambda(r,t)} \frac{(\Lambda(r,t))^k}{k!} \\
&= e^{-\Lambda(r,t)} \int_r^t \frac{(\Lambda(r,s))^{k-1}}{(k-1)!} \lambda(s)\, ds \\
&= \int_r^t e^{-\Lambda(r,s)} \frac{(\Lambda(r,s))^{k-1}}{(k-1)!} \lambda(s)\, e^{-\Lambda(s,t)}\, ds \\
&= \int_r^t p_{n,n+k-1}(r,s)\, \lambda(s)\, p_{n+k,n+k}(s,t)\, ds \ .
\end{aligned}
$$

It now follows from Theorem 3.2.1 that $\{N_t\}_{t\in\mathbf{R}_+}$ is regular with intensities $\{\lambda_n\}_{n\in\mathbf{N}}$ satisfying $\lambda_n(t) = \lambda(t)$ for all $n \in \mathbf{N}$ and $t \in \mathbf{R}_+$. Therefore, (c) implies (b).
• Assume finally that (b) holds. Since $\{N_t\}_{t\in\mathbf{R}_+}$ has independent increments, it follows from Theorem 3.1.1 that $\{N_t\}_{t\in\mathbf{R}_+}$ is a Markov process. Therefore, (b) implies (a).   □

The following result is a partial generalization of Lemma 2.3.1:

3.3.3 Lemma (Multinomial Criterion). *If the claim number process $\{N_t\}_{t\in\mathbf{R}_+}$ is an inhomogeneous Poisson process with intensity λ, then the identity*

$$
P\left[\bigcap_{j=1}^m \{N_{t_j} - N_{t_{j-1}} = k_j\}\,\middle|\,\{N_{t_m} = n\}\right] = \frac{n!}{\prod_{j=1}^m k_j!} \prod_{j=1}^m \left(\frac{\int_{t_{j-1}}^{t_j} \lambda(s)\, ds}{\int_0^{t_m} \lambda(s)\, ds}\right)^{k_j}
$$

holds for all $m \in \mathbf{N}$ and $t_0, t_1, \ldots, t_m \in \mathbf{R}_+$ such that $0 = t_0 < t_1 < \ldots < t_m$ and for all $n \in \mathbf{N}_0$ and $k_1, \ldots, k_m \in \mathbf{N}_0$ such that $\sum_{j=1}^m k_j = n$.

Proof.   For all $r, t \in \mathbf{R}_+$ satisfying $r \leq t$, define

$$\Lambda(r,t) := \int_r^t \lambda(s)\, ds .$$

Then we have

$$
P\left[\bigcap_{j=1}^m \{N_{t_j} - N_{t_{j-1}} = k_j\} \,\middle|\, \{N_{t_m} = n\}\right] = \frac{P\left[\bigcap_{j=1}^m \{N_{t_j} - N_{t_{j-1}} = k_j\}\right]}{P[\{N_{t_m} = n\}]}
$$

$$
= \frac{\prod_{j=1}^m P[\{N_{t_j} - N_{t_{j-1}} = k_j\}]}{P[\{N_{t_m} = n\}]}
$$

$$
= \frac{\prod_{j=1}^m e^{-\Lambda(t_{j-1},t_j)} \dfrac{(\Lambda(t_{j-1},t_j))^{k_j}}{k_j!}}{e^{-\Lambda(0,t_m)} \dfrac{(\Lambda(0,t_m))^n}{n!}}
$$

$$
= \frac{n!}{\prod_{j=1}^m k_j!} \prod_{j=1}^m \left(\frac{\Lambda(t_{j-1},t_j)}{\Lambda(0,t_m)}\right)^{k_j} ,
$$

as was to be shown. $\square$

Problems

3.3.A   Assume that the claim number process has independent increments and is regular. Then its intensities are all identical.

3.3.B   The following are equivalent:
(a) The claim number process is a regular Markov process and its intensities are all identical.
(b) The claim number process has independent increments and is regular.
(c) The claim number process is an inhomogeneous Poisson process.

3.3.C   Consider a continuous function $\lambda : \mathbf{R}_+ \to (0,\infty)$ satisfying $\int_0^\infty \lambda(s)\, ds = \infty$. For all $t \in \mathbf{R}_+$, define

$$\varrho(t) := \int_0^t \lambda(s)\, ds$$

as well as

$$N_t^\flat := N_{\varrho(t)}$$

and

$$N_t^\natural := N_{\varrho^{-1}(t)} .$$

(a) If the claim number process $\{N_t\}_{t \in \mathbf{R}_+}$ is a Poisson process with parameter 1, then $\{N_t^\flat\}_{t \in \mathbf{R}_+}$ is an inhomogeneous Poisson process with intensity λ.
(b) If the claim number process $\{N_t\}_{t \in \mathbf{R}_+}$ is an inhomogeneous Poisson process with intensity λ, then $\{N_t^\natural\}_{t \in \mathbf{R}_+}$ is a Poisson process with parameter 1.

3.3.D Consider a continuous function $\lambda : \mathbf{R}_+ \to (0, \infty)$ satisfying $\int_0^\infty \lambda(s)\,ds = \infty$. For all $t \in \mathbf{R}_+$, define

$$\varrho(t) := \int_0^t \lambda(s)\,ds\ .$$

Then the following are equivalent:
(a) The claim number process $\{N_t\}_{t \in \mathbf{R}_+}$ is an inhomogeneous Poisson process with intensity λ.
(b) The claim number process $\{N_t\}_{t \in \mathbf{R}_+}$ has independent increments and satisfies $E[N_t] = \varrho(t)$ for all $t \in \mathbf{R}_+$.
(c) The process $\{N_t - \varrho(t)\}_{t \in \mathbf{R}_+}$ is a martingale.

3.3.E **Prediction:** If the claim number process $\{N_t\}_{t \in \mathbf{R}_+}$ is an inhomogeneous Poisson process with intensity λ, then the inequality

$$E\left[\left(N_{t+h} - \left(N_t + \int_t^{t+h} \lambda(s)\,ds\right)\right)^2\right] \le E[(N_{t+h} - Z)^2]$$

holds for all $t, h \in \mathbf{R}_+$ and for every random variable Z satisfying $E[Z^2] < \infty$ and $\sigma(Z) \subseteq \mathcal{F}_t$.

3.4   A Characterization of Homogeneity

In the present section, we study claim number processes which are regular Markov processes with intensities which are all constant.

3.4.1 Lemma.   *Assume that the claim number process $\{N_t\}_{t \in \mathbf{R}_+}$ satisfies the Chapman-Kolmogorov equations and is regular. Then the following are equivalent:*
(a) $\{N_t\}_{t \in \mathbf{R}_+}$ *is homogeneous.*
(b) *The intensities of $\{N_t\}_{t \in \mathbf{R}_+}$ are all constant.*

Proof.   • Assume first that (a) holds and consider $n \in \mathbf{N}_0$. For all $s, t \in (0, \infty)$, we have

$$\begin{aligned}
\lambda_{n+1}(s) &= \lim_{h \to 0} \frac{1}{h} p_{n,n+1}(s, s+h) \\
&= \lim_{h \to 0} \frac{1}{h} p_{n,n+1}(t, t+h) \\
&= \lambda_{n+1}(t)\ .
\end{aligned}$$

Thus, λ_{n+1} is constant on $(0, \infty)$ and hence, by continuity, on $\mathbf{R}_+$. Therefore, (a) implies (b).
• Assume now that (b) holds and consider a sequence $\{\alpha_n\}_{n \in \mathbf{N}}$ in $(0, \infty)$ such that $\lambda_n(t) = \alpha_n$ holds for all $n \in \mathbf{N}$ and $t \in \mathbf{R}_+$.

(1) For all $n \in \mathbf{N}_0$ and $s, t, h \in \mathbf{R}_+$ such that (n, s) and (n, t) are admissible, we have

$$
\begin{aligned}
p_{n,n}(s, s+h) &= e^{-\int_s^{s+h} \lambda_{n+1}(u)\, du} \\
&= e^{-\int_s^{s+h} \alpha_{n+1}\, du} \\
&= e^{-\alpha_{n+1} h} ,
\end{aligned}
$$

and hence

$$
p_{n,n}(s, s+h) = p_{n,n}(t, t+h) .
$$

(2) Assume now that the identity

$$
p_{n,n+k}(s, s+h) = p_{n,n+k}(t, t+h)
$$

holds for some $k \in \mathbf{N}_0$ and for all $n \in \mathbf{N}_0$ and $s, t, h \in \mathbf{R}_+$ such that (n, s) and (n, t) are admissible (which because of (1) is the case for $k = 0$). Then we have

$$
\begin{aligned}
p_{n,n+k+1}(s, s+h) &= \int_s^{s+h} p_{n,n+k}(s, u)\, \lambda_{n+k+1}(u)\, p_{n+k+1,n+k+1}(u, s+h)\, du \\
&= \int_s^{s+h} p_{n,n+k}(t, t-s+u)\, \alpha_{n+k+1}\, p_{n+k+1,n+k+1}(t-s+u, t+h)\, du \\
&= \int_t^{t+h} p_{n,n+k}(t, v)\, \alpha_{n+k+1}\, p_{n+k+1,n+k+1}(v, t+h)\, dv \\
&= \int_t^{t+h} p_{n,n+k}(t, v)\, \lambda_{n+k+1}(v)\, p_{n+k+1,n+k+1}(v, t+h)\, dv \\
&= p_{n,n+k+1}(t, t+h)
\end{aligned}
$$

for all $n \in \mathbf{N}_0$ and $s, t, h \in \mathbf{R}_+$ such that (n, s) and (n, t) are admissible
(3) Because of (1) and (2), (b) implies (a). $\qquad \square$

The main result of this section is the following characterization of homogeneous regular Markov processes:

3.4.2 Theorem. *Let $\{\alpha_n\}_{n \in \mathbf{N}}$ be a sequence of real numbers in $(0, \infty)$. Then the following are equivalent:*
(a) *The claim number process $\{N_t\}_{t \in \mathbf{R}_+}$ is a regular Markov process with intensities $\{\lambda_n\}_{n \in \mathbf{N}}$ satisfying $\lambda_n(t) = \alpha_n$ for all $n \in \mathbf{N}$ and $t \in \mathbf{R}_+$.*
(b) *The sequence of claim interarrival times $\{W_n\}_{n \in \mathbf{N}}$ is independent and satisfies $P_{W_n} = \mathbf{Exp}(\alpha_n)$ for all $n \in \mathbf{N}$.*

Proof. For $n \in \mathbf{N}$, let $\mathbf{T}_n$ and $\mathbf{W}_n$ denote the random vectors $\Omega \to \mathbf{R}^n$ with coordinates T_i and W_i, respectively, and let $\mathbf{M}_n$ denote the $(n \times n)$-matrix with entries

$$
m_{ij} := \begin{cases} 1 & \text{if } i \geq j \\ 0 & \text{if } i < j . \end{cases}
$$

Then $\mathbf{M}_n$ is invertible and satisfies

$$\det \mathbf{M}_n = 1 .$$

Moreover, we have $\mathbf{T}_n = \mathbf{M}_n \circ \mathbf{W}_n$ and hence

$$\mathbf{W}_n = \mathbf{M}_n^{-1} \circ \mathbf{T}_n$$

as well as $\mathbf{T}_n^{-1} = \mathbf{W}_n^{-1} \circ \mathbf{M}_n^{-1}$. Furthermore, let $\mathbf{1}_n$ denote the vector in $\mathbf{R}^n$ with all coordinates being equal to one, and let $\langle ., . \rangle$ denote the inner product on $\mathbf{R}^n$.
• Assume first that (a) holds. Since the claim arrival process is more directly related to the claim number process than the claim interarrival process is, we shall first determine the finite dimensional distributions of the claim arrival process and then apply the identity $\mathbf{W}_n = \mathbf{M}_n^{-1} \circ \mathbf{T}_n$ to obtain the finite dimensional distributions of the claim interarrival process.
Consider two sequences $\{r_j\}_{j\in\mathbf{N}}$ and $\{t_j\}_{j\in\mathbf{N}_0}$ of real numbers satisfying $t_0 = 0$ and $t_{j-1} \le r_j < t_j$ for all $j \in \mathbf{N}$. We first exploit regularity and then the Markov property in order to determine the finite dimensional distributions of the claim arrival process.
(1) *For each admissible pair (j,r) and all $t \in [r,\infty)$, we have*

$$p_{j,j}(r,t) = e^{-\alpha_{j+1}(t-r)} .$$

Indeed, regularity yields

$$
\begin{aligned}
p_{j,j}(r,t) &= e^{-\int_r^t \lambda_{j+1}(s)\,ds} \\
&= e^{-\int_r^t \alpha_{j+1}\,ds} \\
&= e^{-\alpha_{j+1}(t-r)} .
\end{aligned}
$$

(2) *For all $j \in \mathbf{N}$ and $r,t \in (0,\infty)$ such that $r \le t$, we have*

$$
p_{j-1,j}(r,t) =
\begin{cases}
\dfrac{\alpha_j}{\alpha_{j+1}-\alpha_j}\left(e^{-\alpha_j(t-r)} - e^{-\alpha_{j+1}(t-r)}\right) & \text{if } \alpha_j \ne \alpha_{j+1} \\[2ex]
\alpha_j(t-r)\,e^{-\alpha_j(t-r)} & \text{if } \alpha_j = \alpha_{j+1} .
\end{cases}
$$

Indeed, if $\alpha_j \ne \alpha_{j+1}$, then regularity together with (1) yields

$$
\begin{aligned}
p_{j-1,j}(r,t) &= \int_r^t p_{j-1,j-1}(r,s)\,\lambda_j(s)\,p_{j,j}(s,t)\,ds \\
&= \int_r^t e^{-\alpha_j(s-r)}\,\alpha_j\,e^{-\alpha_{j+1}(t-s)}\,ds \\
&= \alpha_j\,e^{(\alpha_j r - \alpha_{j+1} t)} \int_r^t e^{(\alpha_{j+1}-\alpha_j)s}\,ds \\
&= \alpha_j\,e^{(\alpha_j r - \alpha_{j+1} t)}\,\frac{1}{\alpha_{j+1}-\alpha_j}\left(e^{(\alpha_{j+1}-\alpha_j)t} - e^{(\alpha_{j+1}-\alpha_j)r}\right) \\
&= \frac{\alpha_j}{\alpha_{j+1}-\alpha_j}\left(e^{-\alpha_j(t-r)} - e^{-\alpha_{j+1}(t-r)}\right) ;
\end{aligned}
$$

similarly, if $\alpha_j = \alpha_{j+1}$, then

$$
\begin{aligned}
p_{j-1,j}(r,t) &= \int_r^t p_{j-1,j-1}(r,s)\,\lambda_j(s)\,p_{j,j}(s,t)\,ds \\
&= \int_r^t e^{-\alpha_j(s-r)}\,\alpha_j\,e^{-\alpha_{j+1}(t-s)}\,ds \\
&= \int_r^t e^{-\alpha_j(s-r)}\,\alpha_j\,e^{-\alpha_j(t-s)}\,ds \\
&= \int_r^t \alpha_j\,e^{-\alpha_j(t-r)}\,ds \\
&= \alpha_j(t-r)\,e^{-\alpha_j(t-r)}\,.
\end{aligned}
$$

(3) *For all $j \in \mathbf{N}$ and $h \in (0,\infty)$, we have*

$$
\begin{aligned}
& p_{j-1,j-1}(h,h+r_j)\,p_{j-1,j}(r_j,t_j)\,p_{j,j}(t_j,r_{j+1}) \\
&= \int_{r_j}^{t_j} \alpha_j\,e^{(\alpha_{j+1}-\alpha_j)s_j}\,ds_j \cdot p_{j,j}(h,h+r_{j+1})\,.
\end{aligned}
$$

Indeed, if $\alpha_j \neq \alpha_{j+1}$, then (1) and (2) yield

$$
\begin{aligned}
& p_{j-1,j-1}(h,h+r_j)\,p_{j-1,j}(r_j,t_j)\,p_{j,j}(t_j,r_{j+1}) \\
&= e^{-\alpha_j r_j} \cdot \frac{\alpha_j}{\alpha_{j+1}-\alpha_j}\left(e^{-\alpha_j(t_j-r_j)} - e^{-\alpha_{j+1}(t_j-r_j)}\right)\cdot e^{-\alpha_{j+1}(r_{j+1}-t_j)} \\
&= \frac{\alpha_j}{\alpha_{j+1}-\alpha_j}\left(e^{(\alpha_{j+1}-\alpha_j)t_j} - e^{(\alpha_{j+1}-\alpha_j)r_j}\right)\cdot e^{-\alpha_{j+1}r_{j+1}} \\
&= \int_{r_j}^{t_j} \alpha_j\,e^{(\alpha_{j+1}-\alpha_j)s_j}\,ds_j \cdot p_{j,j}(h,h+r_{j+1})\,;
\end{aligned}
$$

similarly, if $\alpha_j = \alpha_{j+1}$, then

$$
\begin{aligned}
& p_{j-1,j-1}(h,h+r_j)\,p_{j-1,j}(r_j,t_j)\,p_{j,j}(t_j,r_{j+1}) \\
&= e^{-\alpha_j r_j} \cdot \alpha_j(t_j-r_j)\,e^{-\alpha_j(t_j-r_j)}\cdot e^{-\alpha_{j+1}(r_{j+1}-t_j)} \\
&= e^{-\alpha_j r_j} \cdot \alpha_j(t_j-r_j)\,e^{-\alpha_j(t_j-r_j)}\cdot e^{-\alpha_j(r_{j+1}-t_j)} \\
&= \alpha_j(t_j-r_j)\cdot e^{-\alpha_j r_{j+1}} \\
&= \int_{r_j}^{t_j} \alpha_j\,ds_j \cdot p_{j,j}(h,h+r_{j+1}) \\
&= \int_{r_j}^{t_j} \alpha_j\,e^{(\alpha_{j+1}-\alpha_j)s_j}\,ds_j \cdot p_{j,j}(h,h+r_{j+1})\,.
\end{aligned}
$$

(4) *For all $n \in \mathbf{N}$, we have*

$$P\left[\{N_{r_n} = n-1\} \cap \bigcap_{j=1}^{n-1}\{N_{t_j} = j\} \cap \{N_{r_j} = j-1\}\right] > 0$$

and

$$P\left[\bigcap_{j=1}^{n}\{N_{t_j} = j\} \cap \{N_{r_j} = j-1\}\right] > 0 .$$

This follows by induction, using (1) and (2) and the Markov property:
For $n = 1$, we have

$$\begin{aligned}
P[\{N_{r_1} = 0\}] &= P[\{N_{r_1} = 0\}|\{N_0 = 0\}] \\
&= p_{0,0}(0, r_1) \\
&> 0
\end{aligned}$$

and

$$\begin{aligned}
P[\{N_{t_1} = 1\} \cap \{N_{r_1} = 0\}] &= P[\{N_{t_1} = 1\}|\{N_{r_1} = 0\}] \cdot P[\{N_{r_1} = 0\}] \\
&= p_{0,0}(0, r_1)\, p_{0,1}(r_1, t_1) \\
&> 0 .
\end{aligned}$$

Assume now that the assertion holds for some $n \in \mathbf{N}$. Then we have

$$\begin{aligned}
&P\left[\{N_{r_{n+1}} = n\} \cap \bigcap_{j=1}^{n}\{N_{t_j} = j\} \cap \{N_{r_j} = j-1\}\right] \\
&= P\left[\{N_{r_{n+1}} = n\} \middle| \bigcap_{j=1}^{n}\{N_{t_j} = j\} \cap \{N_{r_j} = j-1\}\right] \\
&\qquad \cdot P\left[\bigcap_{j=1}^{n}\{N_{t_j} = j\} \cap \{N_{r_j} = j-1\}\right] \\
&= P[\{N_{r_{n+1}} = n\}|\{N_{t_n} = n\}] \\
&\qquad \cdot P\left[\bigcap_{j=1}^{n}\{N_{t_j} = j\} \cap \{N_{r_j} = j-1\}\right] \\
&= p_{n,n}(t_n, r_{n+1}) \cdot P\left[\bigcap_{j=1}^{n}\{N_{t_j} = j\} \cap \{N_{r_j} = j-1\}\right] \\
&> 0
\end{aligned}$$

and

$$P\left[\bigcap_{j=1}^{n+1}\{N_{t_j} = j\}\cap\{N_{r_j} = j-1\}\right]$$

$$= P\left[\{N_{t_{n+1}} = n+1\}\Bigg|\{N_{r_{n+1}} = n\}\cap\bigcap_{j=1}^{n}\{N_{t_j} = j\}\cap\{N_{r_j} = j-1\}\right]$$

$$\cdot P\left[\{N_{r_{n+1}} = n\}\cap\bigcap_{j=1}^{n}\{N_{t_j} = j\}\cap\{N_{r_j} = j-1\}\right]$$

$$= P[\{N_{t_{n+1}} = n+1\}|\{N_{r_{n+1}} = n\}]$$

$$\cdot P\left[\{N_{r_{n+1}} = n\}\cap\bigcap_{j=1}^{n}\{N_{t_j} = j\}\cap\{N_{r_j} = j-1\}\right]$$

$$= p_{n,n+1}(r_{n+1},t_{n+1})\cdot P\left[\{N_{r_{n+1}} = n\}\cap\bigcap_{j=1}^{n}\{N_{t_j} = j\}\cap\{N_{r_j} = j-1\}\right]$$

$$> 0.$$

(5) *For all $n \in \mathbf{N}$, we have*

$$P\left[\bigcap_{j=1}^{n}\{r_j < T_j \leq t_j\}\right] = \prod_{j=1}^{n-1}\int_{r_j}^{t_j}\alpha_j e^{(\alpha_{j+1}-\alpha_j)s_j}\,ds_j \cdot \int_{r_n}^{t_n}\alpha_n e^{\alpha_n s_n}\,ds_n\,.$$

Indeed, using (4), the Markov property, (3), and (1), we obtain, for all $h \in (0.\infty)$,

$$P\left[\bigcap_{j=1}^{n}\{r_j < T_j \leq t_j\}\right]$$

$$= P\left[\bigcap_{j=1}^{n}\{N_{r_j} < j \leq N_{t_j}\}\right]$$

$$= P\left[\{N_{t_n} \geq n\}\cap\{N_{r_n} = n-1\}\cap\bigcap_{j=1}^{n-1}\{N_{t_j} = j\}\cap\{N_{r_j} = j-1\}\right]$$

$$= P[\{N_{t_n} \geq n\}|\{N_{r_n} = n-1\}]$$

$$\cdot \prod_{j=1}^{n-1}\left(P[\{N_{r_{j+1}} = j\}|\{N_{t_j} = j\}]\cdot P[\{N_{t_j} = j\}|\{N_{r_j} = j-1\}]\right)$$

$$\cdot P[\{N_{r_1} = 0\}|\{N_0 = 0\}]$$

$$= p_{0,0}(0,r_1)\cdot\prod_{j=1}^{n-1}p_{j-1,j}(r_j,t_j)\,p_{j,j}(t_j,r_{j+1})\cdot\left(1 - p_{n-1,n-1}(r_n,t_n)\right)$$

$$
= p_{0,0}(h, h+r_1) \cdot \prod_{j=1}^{n-1} p_{j-1,j}(r_j, t_j)\, p_{j,j}(t_j, r_{j+1}) \cdot \left(1 - p_{n-1,n-1}(r_n, t_n)\right)
$$

$$
= \prod_{j=1}^{n-1} \int_{r_j}^{t_j} \alpha_j\, e^{(\alpha_{j+1}-\alpha_j)s_j}\, ds_j \cdot p_{n-1,n-1}(h, h+r_n) \cdot \left(1 - p_{n-1,n-1}(r_n, t_n)\right)
$$

$$
= \prod_{j=1}^{n-1} \int_{r_j}^{t_j} \alpha_j\, e^{(\alpha_{j+1}-\alpha_j)s_j}\, ds_j \cdot e^{-\alpha_n r_n} \cdot \left(1 - e^{-\alpha_n(t_n-r_n)}\right)
$$

$$
= \prod_{j=1}^{n-1} \int_{r_j}^{t_j} \alpha_j\, e^{(\alpha_{j+1}-\alpha_j)s_j}\, ds_j \cdot \left(e^{-\alpha_n r_n} - e^{-\alpha_n t_n}\right)
$$

$$
= \prod_{j=1}^{n-1} \int_{r_j}^{t_j} \alpha_j\, e^{(\alpha_{j+1}-\alpha_j)s_j}\, ds_j \cdot \int_{r_n}^{t_n} \alpha_n\, e^{-\alpha_n s_n}\, ds_n \ .
$$

(6) Consider $n \in \mathbf{N}$ and define $\mathbf{f}_n : \mathbf{R}^n \to \mathbf{R}_+$ by letting

$$
\mathbf{f}_n(\mathbf{w}) \ := \ \prod_{j=1}^{n} \alpha_j\, e^{-\alpha_j w_j} \chi_{(0,\infty)}(w_j) \ .
$$

Define $s_0 := 0$ and $A := \bigtimes_{j=1}^{n}(r_j, t_j]$. Because of (5), we obtain

$$
P[\{\mathbf{T}_n \in A\}] \ = \ P\left[\bigcap_{j=1}^{n}\{r_j < T_j \le t_j\}\right]
$$

$$
= \ \prod_{j=1}^{n-1} \int_{r_j}^{t_j} \alpha_j\, e^{(\alpha_{j+1}-\alpha_j)s_j}\, ds_j \cdot \int_{r_n}^{t_n} \alpha_n\, e^{-\alpha_n s_n}\, ds_n
$$

$$
= \ \prod_{j=1}^{n-1} \int_{(r_j, t_j]} \alpha_j\, e^{(\alpha_{j+1}-\alpha_j)s_j}\, d\lambda(s_j) \cdot \int_{(r_n, t_n]} \alpha_n\, e^{-\alpha_n s_n}\, d\lambda(s_n)
$$

$$
= \ \int_A \left(\prod_{j=1}^{n-1} \alpha_j\, e^{(\alpha_{j+1}-\alpha_j)s_j}\right) \alpha_n\, e^{-\alpha_n s_n}\, d\lambda^n(\mathbf{s})
$$

$$
= \ \int_A \left(\prod_{j=1}^{n} \alpha_j\, e^{-\alpha_j(s_j-s_{j-1})} \chi_{(0,\infty)}(s_j - s_{j-1})\right) d\lambda^n(\mathbf{s})
$$

$$
= \ \int_A \mathbf{f}_n(\mathbf{M}_n^{-1}(\mathbf{s}))\, d\lambda^n(\mathbf{s}) \ .
$$

(7) Since the sequence $\{T_n\}_{n \in \mathbf{N}_0}$ is strictly increasing with $T_0 = 0$, it follows from (6) that the identity

$$
P[\{\mathbf{T}_n \in A\}] \ = \ \int_A \mathbf{f}_n(\mathbf{M}_n^{-1}(\mathbf{s}))\, d\lambda^n(\mathbf{s})
$$

holds for all $A \in \mathcal{B}(\mathbf{R}^n)$.

(8) Consider $B_1, \ldots, B_n \in \mathcal{B}(\mathbf{R})$ and let $B := \times_{j=1}^{n} B_j$. Since $\mathbf{W}_n = \mathbf{M}_n^{-1} \circ \mathbf{T}_n$, the identity established in (7) yields

$$
\begin{aligned}
P_{\mathbf{W}_n}[B] &= P[\{\mathbf{W}_n \in B\}] \\
&= P[\{\mathbf{M}_n^{-1} \circ \mathbf{T}_n \in B\}] \\
&= P[\{\mathbf{T}_n \in \mathbf{M}_n(B)\}] \\
&= \int_{\mathbf{M}_n(B)} \mathbf{f}_n(\mathbf{M}_n^{-1}(\mathbf{s})) \, d\lambda^n(\mathbf{s}) \\
&= \int_B \mathbf{f}_n(\mathbf{w}) \, d\lambda^n_{\mathbf{M}_n^{-1}}(\mathbf{w}) \\
&= \int_B \mathbf{f}_n(\mathbf{w}) \frac{1}{|\det \mathbf{M}_n^{-1}|} \, d\lambda^n(\mathbf{w}) \\
&= \int_B \mathbf{f}_n(\mathbf{w}) \, d\lambda^n(\mathbf{w}) \\
&= \int_B \left(\prod_{j=1}^{n} \alpha_j \, e^{-\alpha_j w_j} \chi_{(0,\infty)}(w_j) \right) d\lambda^n(\mathbf{w}) \\
&= \prod_{j=1}^{n} \int_{B_j} \alpha_j \, e^{-\alpha_j w_j} \chi_{(0,\infty)}(w_j) \, d\lambda(w_j) \, .
\end{aligned}
$$

(9) Because of (8), the sequence $\{W_n\}_{n \in \mathbf{N}}$ is independent and satisfies

$$
P_{W_n} = \mathbf{Exp}(\alpha_n)
$$

for all $n \in \mathbf{N}$. Therefore, (a) implies (b).

• Assume now that (b) holds.

(1) To establish the Markov property, consider $m \in \mathbf{N}$, $t_1, \ldots, t_m, t_{m+1} \in (0, \infty)$, and $k_1, \ldots, k_m \in \mathbf{N}_0$ such that $t_1 < \ldots < t_m < t_{m+1}$ and $P[\bigcap_{j=1}^{m} \{N_{t_j} = k_j\}] > 0$. In the case where $k_m = 0$, it is clear that the identity

$$
P\left[\{N_{t_{m+1}} = k_{m+1}\} \Big| \bigcap_{j=1}^{m} \{N_{t_j} = k_j\} \right] = P[\{N_{t_{m+1}} = l\} | \{N_{t_m} = k_m\}]
$$

holds for all $k_{m+1} \in \mathbf{N}_0$ such that $k_m \leq k_{m+1}$.

In the case where $k_m \in \mathbf{N}$, define $k_0 := 0$ and $n := k_m$, and let $l \in \{0, 1, \ldots, m-1\}$ be the unique integer satisfying $k_l < k_{l+1} = k_m = n$. Then there exists a rectangle $B \subseteq \times_{j=1}^{n} (0, t_{l+1}]$ such that

$$
\left(\bigcap_{j=1}^{l} \{T_{k_j} \leq t_j < T_{k_j+1}\} \right) \cap \{T_n \leq t_{l+1}\} = \mathbf{T}_n^{-1}(B) \, .
$$

Letting $A := \mathbf{M}_n^{-1}(B)$, we obtain

$$
\left(\bigcap_{j=1}^{l}\{T_{k_j} \leq t_j < T_{k_j+1}\}\right) \cap \{T_n \leq t_{l+1}\} = T_n^{-1}(B)
$$
$$
= \mathbf{W}_n^{-1}(\mathbf{M}_n^{-1}(B))
$$
$$
= \mathbf{W}_n^{-1}(A) .
$$

Using independence of $\mathbf{W}_n$ and W_{n+1}, the transformation formula for integrals, and Fubini's theorem, we obtain

$$
P\left[\bigcap_{j=1}^{m}\{N_{t_j} = k_j\}\right] = P\left[\left(\bigcap_{j=1}^{l}\{N_{t_j} = k_j\}\right) \cap \{N_{t_{l+1}} = n\} \cap \{N_{t_m} = n\}\right]
$$
$$
= P\left[\left(\bigcap_{j=1}^{l}\{T_{k_j} \leq t_j < T_{k_j+1}\}\right) \cap \{T_n \leq t_{l+1}\} \cap \{t_m < T_{n+1}\}\right]
$$
$$
= P[\mathbf{W}_n^{-1}(A) \cap \{t_m - T_n < W_{n+1}\}]
$$
$$
= P[\mathbf{W}_n^{-1}(A) \cap \{t_m - \langle \mathbf{1}_n, \mathbf{W}_n\rangle < W_{n+1}\}]
$$
$$
= \int_A \left(\int_{(t_m - \langle \mathbf{1}_n, \mathbf{s}\rangle, \infty)} dP_{W_{n+1}}(w)\right) dP_{\mathbf{W}_n}(\mathbf{s})
$$
$$
= \int_A \left(\int_{(t_m - \langle \mathbf{1}_n, \mathbf{s}\rangle, \infty)} \alpha_{n+1}\, e^{-\alpha_{n+1} w}\, d\lambda(w)\right) dP_{\mathbf{W}_n}(\mathbf{s})
$$
$$
= \int_A e^{\alpha_{n+1}\langle \mathbf{1}_n, \mathbf{s}\rangle}\left(\int_{(t_m, \infty)} \alpha_{n+1}\, e^{-\alpha_{n+1} v}\, d\lambda(v)\right) dP_{\mathbf{W}_n}(\mathbf{s})
$$
$$
= \int_A e^{\alpha_{n+1}\langle \mathbf{1}_n, \mathbf{s}\rangle}\, dP_{\mathbf{W}_n}(\mathbf{s}) \cdot \int_{(t_m, \infty)} dP_{W_{n+1}}(v) .
$$

Also, using the same arguments as before, we obtain

$$
P[\{N_{t_m} = k_m\}] = P[\{N_{t_m} = n\}]
$$
$$
= P[\{T_n \leq t_m < T_{n+1}\}]
$$
$$
= P[\{T_n \leq t_m\} \cap \{t_m - T_n < W_{n+1}\}]
$$
$$
= \int_{(-\infty, t_m]} \left(\int_{(t_m - s, \infty)} dP_{W_{n+1}}(w)\right) dP_{T_n}(s)
$$
$$
= \int_{(-\infty, t_m]} \left(\int_{(t_m - s, \infty)} \alpha_{n+1}\, e^{-\alpha_{n+1} w}\, d\lambda(w)\right) dP_{T_n}(s)
$$
$$
= \int_{(-\infty, t_m]} e^{\alpha_{n+1} s}\left(\int_{(t_m, \infty)} \alpha_{n+1}\, e^{-\alpha_{n+1} v}\, d\lambda(v)\right) dP_{T_n}(s)
$$
$$
= \int_{(-\infty, t_m]} e^{\alpha_{n+1} s}\, dP_{T_n}(s) \cdot \int_{(t_m, \infty)} dP_{W_{n+1}}(v) .
$$

Consider now $k \in \mathbf{N}_0$ such that $k_m < k$, and define $U := T_k - T_{n+1} = T_k - T_n - W_{n+1}$. Then we have

$$
P\left[\{N_{t_{m+1}} \geq k\} \cap \bigcap_{j=1}^{m}\{N_{t_j} = k_j\}\right]
$$

$$
= P\left[\left(\bigcap_{j=1}^{l}\{N_{t_j} = k_j\}\right) \cap \{N_{t_{l+1}} = n\} \cap \{N_{t_m} = n\} \cap \{N_{t_{m+1}} \geq k\}\right]
$$

$$
= P\left[\left(\bigcap_{j=1}^{l}\{T_{k_j} \leq t_j < T_{k_j+1}\}\right) \cap \{T_n \leq t_{l+1}\} \cap \{t_m < T_{n+1}\} \cap \{T_k \leq t_{m+1}\}\right]
$$

$$
= P[\mathbf{W}_n^{-1}(A) \cap \{t_m - T_n < W_{n+1}\} \cap \{U \leq t_{m+1} - T_n - W_{n+1}\}]
$$

$$
= P[\mathbf{W}_n^{-1}(A) \cap \{t_m - \langle \mathbf{1}_n, \mathbf{W}_n \rangle < W_{n+1}\} \cap \{U \leq t_{m+1} - \langle \mathbf{1}_n, \mathbf{W}_n \rangle - W_{n+1}\}]
$$

$$
= \int_A \left(\int_{(t_m - \langle \mathbf{1}_n, \mathbf{s} \rangle, \infty)} \left(\int_{(-\infty, t_{m+1} - \langle \mathbf{1}_n, \mathbf{s} \rangle - w]} dP_U(u)\right) dP_{W_{n+1}}(w)\right) dP_{\mathbf{W}_n}(\mathbf{s})
$$

$$
= \int_A \left(\int_{(t_m - \langle \mathbf{1}_n, \mathbf{s} \rangle, \infty)} P[\{U \leq t_{m+1} - \langle \mathbf{1}_n, \mathbf{s} \rangle - w\}] dP_{W_{n+1}}(w)\right) dP_{\mathbf{W}_n}(\mathbf{s})
$$

$$
= \int_A \left(\int_{(t_m - \langle \mathbf{1}_n, \mathbf{s} \rangle, \infty)} P[\{U \leq t_{m+1} - \langle \mathbf{1}_n, \mathbf{s} \rangle - w\}] \alpha_{n+1} e^{-\alpha_{n+1} w} d\lambda(w)\right) dP_{\mathbf{W}_n}(\mathbf{s})
$$

$$
= \int_A e^{\alpha_{n+1} \langle \mathbf{1}_n, \mathbf{s} \rangle} \left(\int_{(t_m, \infty)} P[\{U \leq t_{m+1} - v\}] \alpha_{n+1} e^{-\alpha_{n+1} v} d\lambda(v)\right) dP_{\mathbf{W}_n}(\mathbf{s})
$$

$$
= \int_A e^{\alpha_{n+1} \langle \mathbf{1}_n, \mathbf{s} \rangle} dP_{\mathbf{W}_n}(\mathbf{s}) \cdot \int_{(t_m, \infty)} P[\{U \leq t_{m+1} - v\}] dP_{W_{n+1}}(v)
$$

as well as

$$
P[\{N_{t_{m+1}} \geq k\} \cap \{N_{t_m} = k_m\}]
$$

$$
= P[\{N_{t_{m+1}} \geq k\} \cap \{N_{t_m} = n\}]
$$

$$
= P[\{T_n \leq t_m < T_{n+1}\} \cap \{T_k \leq t_{m+1}\}]
$$

$$
= P[\{T_n \leq t_m\} \cap \{t_m - T_n < W_{n+1}\} \cap \{U \leq t_{m+1} - T_n - W_{n+1}\}]
$$

$$
= \int_{(-\infty, t_m]} \left(\int_{(t_m - s, \infty)} \left(\int_{(-\infty, t_{m+1} - s - w]} dP_U(u)\right) dP_{W_{n+1}}(w)\right) dP_{T_n}(s)
$$

$$
= \int_{(-\infty, t_m]} \left(\int_{(t_m - s, \infty)} P[\{U \leq t_{m+1} - s - w\}] dP_{W_{n+1}}(w)\right) dP_{T_n}(s)
$$

$$
= \int_{(-\infty, t_m]} \left(\int_{(t_m - s, \infty)} P[\{U \leq t_{m+1} - s - w\}] \alpha_{n+1} e^{-\alpha_{n+1} w} d\lambda(w)\right) dP_{T_n}(s)
$$

$$
= \int_{(-\infty, t_m]} e^{\alpha_{n+1} s} \left(\int_{(t_m, \infty)} P[\{U \leq t_{m+1} - v\}] \alpha_{n+1} e^{-\alpha_{n+1} v} d\lambda(v)\right) dP_{T_n}(s)
$$

$$
= \int_{(-\infty, t_m]} e^{\alpha_{n+1} s} dP_{T_n}(s) \cdot \int_{(t_m, \infty)} P[\{U \leq t_{m+1} - v\}] dP_{W_{n+1}}.
$$

This yields

$$P\left[\{N_{t_{m+1}} \geq k\} \middle| \bigcap_{j=1}^{m}\{N_{t_j} = k_j\}\right]$$

$$= \frac{P\left[\{N_{t_{m+1}} \geq k\} \cap \bigcap_{j=1}^{m}\{N_{t_j} = k_j\}\right]}{P\left[\bigcap_{j=1}^{m}\{N_{t_j} = k_j\}\right]}$$

$$= \frac{\int_A e^{\alpha_{n+1}\langle 1_n, \mathbf{s}\rangle}\, dP_{\mathbf{W}_n}(\mathbf{s}) \cdot \int_{(t_m,\infty)} P[\{U \leq t_{m+1} - v\}]\, dP_{W_{n+1}}(v)}{\int_A e^{\alpha_{n+1}\langle 1_n, \mathbf{s}\rangle}\, dP_{\mathbf{W}_n}(\mathbf{s}) \cdot \int_{(t_m,\infty)} dP_{W_{n+1}}(v)}$$

$$= \frac{\int_{(t_m,\infty)} P[\{U \leq t_{m+1} - v\}]\, dP_{W_{n+1}}(v)}{\int_{(t_m,\infty)} dP_{W_{n+1}}(v)}$$

$$= \frac{\int_{(-\infty,t_m]} e^{\alpha_{n+1}s}\, dP_{T_n}(s) \cdot \int_{(t_m,\infty)} P[\{U \leq t_{m+1} - v\}]\, dP_{W_{n+1}}(v)}{\int_{(-\infty,t_m]} e^{\alpha_{n+1}s}\, dP_{T_n}(s) \cdot \int_{(t_m,\infty)} dP_{W_{n+1}}(v)}$$

$$= \frac{P[\{N_{t_{m+1}} \geq k\} \cap \{N_{t_m} = k_m\}]}{P[\{N_{t_m} = k_m\}]}$$

$$= P[\{N_{t_{m+1}} \geq k\}|\{N_{t_m} = k_m\}] \,.$$

Therefore, we have

$$P\left[\{N_{t_{m+1}} \geq k\} \middle| \bigcap_{j=1}^{m}\{N_{t_j} = k_j\}\right] = P[\{N_{t_{m+1}} \geq k\}|\{N_{t_m} = k_m\}]$$

for all $k \in \mathbf{N}_0$ such that $k_m < k$.
Of course, the previous identity is also valid if $k_m = k$, and it thus holds for all $k \in \mathbf{N}_0$ such that $k_m \leq k$. But this implies that the identity

$$P\left[\{N_{t_{m+1}} = k_{m+1}\} \middle| \bigcap_{j=1}^{m}\{N_{t_j} = k_j\}\right] = P[\{N_{t_{m+1}} = k_{m+1}\}|\{N_{t_m} = k_m\}]$$

holds for all $k_{m+1} \in \mathbf{N}_0$ such that $k_m \leq k_{m+1}$, which means that $\{N_t\}_{t \in \mathbf{R}_+}$ is a Markov process.

(2) To prove the assertion on regularity, consider an admissible pair (n, t). As before, we obtain

$$
\begin{aligned}
P[\{N_{t+h} = n\} \cap \{N_t = n\}] &= P[\{T_n \leq t\} \cap \{t+h < T_{n+1}\}] \\
&= P[\{T_n \leq t\} \cap \{t+h-T_n < W_{n+1}\}] \\
&= \int_{(-\infty,t]} \left(\int_{(t+h-s,\infty]} dP_{W_{n+1}}(w) \right) dP_{T_n}(s) \\
&= \int_{(-\infty,t]} \left(\int_{(t+h-s,\infty]} \alpha_{n+1}\, e^{-\alpha_{n+1} w}\, d\lambda(w) \right) dP_{T_n}(s) \\
&= \int_{(-\infty,t]} e^{-\alpha_{n+1}(t+h-s)}\, dP_{T_n}(s) \\
&= e^{-\alpha_{n+1}(t+h)} \int_{(-\infty,t]} e^{\alpha_{n+1} s}\, dP_{T_n}(s)
\end{aligned}
$$

for all $h \in \mathbf{R}_+$, hence

$$
P[\{N_t = n\}] = e^{-\alpha_{n+1} t} \int_{(-\infty,t]} e^{\alpha_{n+1} s}\, dP_{T_n}(s)
$$

$$
> 0,
$$

and thus

$$
\begin{aligned}
p_{n,n}(t, t+h) &= P[\{N_{t+h} = n\} | \{N_t = n\}] \\
&= \frac{P[\{N_{t+h} = n\} \cap \{N_t = n\}]}{P[\{N_t = n\}]} \\
&= \frac{e^{-\alpha_{n+1}(t+h)} \displaystyle\int_{(-\infty,t]} e^{\alpha_{n+1} s}\, dP_{T_n}(s)}{e^{-\alpha_{n+1} t} \displaystyle\int_{(-\infty,t]} e^{\alpha_{n+1} s}\, dP_{T_n}(s)} \\
&= e^{-\alpha_{n+1} h}
\end{aligned}
$$

for all $h \in \mathbf{R}_+$.

By what we have shown so far, we have

$$
P[\{N_t = n\}] > 0,
$$

which proves (i).

It is also clear that the function $h \mapsto p_{n,n}(t, t+h)$ is continuous, which proves (ii). Furthermore, we have

$$
\begin{aligned}
\lim_{h \to 0} \frac{1}{h} \left(1 - p_{n,n}(t, t+h) \right) &= \lim_{h \to 0} \frac{1}{h} \left(1 - e^{-\alpha_{n+1} h} \right) \\
&= \alpha_{n+1}.
\end{aligned}
$$

Also, we have

$$P[\{N_{t+h} = n+1\} \cap \{N_t = n\}]$$

$$= P[\{T_n \leq t < T_{n+1} \leq t+h < T_{n+2}\}]$$

$$= P[\{T_n \leq t\} \cap \{t - T_n < W_{n+1} \leq t+h - T_n\} \cap \{t+h - T_n - W_{n+1} < W_{n+2}\}]$$

$$= \int_{(-\infty,t]} \left(\int_{(t-s,t+h-s]} \left(\int_{(t+h-s-w,\infty]} dP_{W_{n+2}}(u) \right) dP_{W_{n+1}}(w) \right) dP_{T_n}(s)$$

$$= \int_{(-\infty,t]} \left(\int_{(t-s,t+h-s]} \left(\int_{(t+h-s-w,\infty]} \alpha_{n+2}\, e^{-\alpha_{n+2}u} \, d\boldsymbol{\lambda}(u) \right) dP_{W_{n+1}}(w) \right) dP_{T_n}(s)$$

$$= \int_{(-\infty,t]} \left(\int_{(t-s,t+h-s]} e^{-\alpha_{n+2}(t+h-s-w)} \, dP_{W_{n+1}}(w) \right) dP_{T_n}(s)$$

$$= \int_{(-\infty,t]} \left(\int_{(t-s,t+h-s]} e^{-\alpha_{n+2}(t+h-s-w)} \, \alpha_{n+1}\, e^{-\alpha_{n+1}w} \, d\boldsymbol{\lambda}(w) \right) dP_{T_n}(s)$$

$$= \int_{(-\infty,t]} \left(\int_{(t,t+h]} e^{-\alpha_{n+2}(t+h-v)} \, \alpha_{n+1}\, e^{-\alpha_{n+1}(v-s)} \, d\boldsymbol{\lambda}(v) \right) dP_{T_n}(s)$$

$$= \alpha_{n+1}\, e^{-\alpha_{n+2}(t+h)} \int_{(-\infty,t]} e^{\alpha_{n+1}s} \, dP_{T_n}(s) \int_{(t,t+h]} e^{(\alpha_{n+2}-\alpha_{n+1})v} \, d\boldsymbol{\lambda}(v) \,,$$

and thus

$$P[\{N_{t+h} = n+1\} | \{N_t = n\}]$$

$$= \frac{P[\{N_{t+h} = n+1\} \cap \{N_t = n\}]}{P[\{N_t = n\}]}$$

$$= \frac{\alpha_{n+1}\, e^{-\alpha_{n+2}(t+h)} \int_{(-\infty,t]} e^{\alpha_{n+1}s} \, dP_{T_n}(s) \int_{(t,t+h]} e^{(\alpha_{n+2}-\alpha_{n+1})v} \, d\boldsymbol{\lambda}(v)}{e^{-\alpha_{n+1}t} \int_{(-\infty,t]} e^{\alpha_{n+1}s} \, dP_{T_n}(s)}$$

$$= \alpha_{n+1}\, e^{\alpha_{n+1}t}\, e^{-\alpha_{n+2}(t+h)} \int_{(t,t+h]} e^{(\alpha_{n+2}-\alpha_{n+1})v} \, d\boldsymbol{\lambda}(v) \,.$$

In the case $\alpha_{n+1} \neq \alpha_{n+2}$, we obtain

$$p_{n,n+1}(t,t+h) = P[\{N_{t+h} = n+1\} | \{N_t = n\}]$$

$$= \alpha_{n+1}\, e^{\alpha_{n+1}t}\, e^{-\alpha_{n+2}(t+h)} \int_{(t,t+h]} e^{(\alpha_{n+2}-\alpha_{n+1})v} \, d\boldsymbol{\lambda}(v)$$

$$= \alpha_{n+1}\, e^{\alpha_{n+1}t}\, e^{-\alpha_{n+2}(t+h)} \frac{e^{(\alpha_{n+2}-\alpha_{n+1})(t+h)} - e^{(\alpha_{n+2}-\alpha_{n+1})t}}{\alpha_{n+2} - \alpha_{n+1}}$$

$$= \frac{\alpha_{n+1}}{\alpha_{n+2} - \alpha_{n+1}} \left(e^{-\alpha_{n+1}h} - e^{-\alpha_{n+2}h} \right) \,,$$

and thus

$$\lim_{h \to 0} \frac{1}{h} p_{n,n+1}(t, t+h) = \lim_{h \to 0} \frac{1}{h} \frac{\alpha_{n+1}}{\alpha_{n+2} - \alpha_{n+1}} \left(e^{-\alpha_{n+1}h} - e^{-\alpha_{n+2}h} \right)$$

$$= \alpha_{n+1};$$

in the case $\alpha_{n+1} = \alpha_{n+2}$, we obtain

$$p_{n,n+1}(t, t+h) = P[\{N_{t+h} = n+1\}|\{N_t = n\}]$$

$$= \alpha_{n+1} e^{\alpha_{n+1}t} e^{-\alpha_{n+2}(t+h)} \int_{(t,t+h]} e^{(\alpha_{n+2} - \alpha_{n+1})v} d\lambda(v)$$

$$= \alpha_{n+1} e^{\alpha_{n+1}t} e^{-\alpha_{n+1}(t+h)} \int_{(t,t+h]} d\lambda(v)$$

$$= \alpha_{n+1} h e^{-\alpha_{n+1}h},$$

and thus

$$\lim_{h \to 0} \frac{1}{h} p_{n,n+1}(t, t+h) = \lim_{h \to 0} \frac{1}{h} \alpha_{n+1} h e^{-\alpha_{n+1}h}$$

$$= \alpha_{n+1}.$$

Thus, in either case we have

$$\lim_{h \to 0} \frac{1}{h} \left(1 - p_{n,n}(t, t+h) \right) = \alpha_{n+1}$$

$$= \lim_{h \to 0} \frac{1}{h} p_{n,n+1}(t, t+h).$$

This proves (iii).
We have thus shown that $\{N_t\}_{t \in \mathbf{R}_+}$ is regular with intensities $\{\lambda_n\}_{n \in \mathbf{N}}$ satisfying $\lambda_n(t) = \alpha_n$ for all $n \in \mathbf{N}$ and $t \in \mathbf{R}_+$. Therefore, (b) implies (a). $\square$

In the situation of Theorem 3.4.2, explosion is either certain or impossible:

3.4.3 Corollary (Zero–One Law on Explosion). *Let $\{\alpha_n\}_{n \in \mathbf{N}}$ be a sequence of real numbers in $(0, \infty)$ and assume that the claim number process $\{N_t\}_{t \in \mathbf{R}_+}$ is a regular Markov process with intensities $\{\lambda_n\}_{n \in \mathbf{N}}$ satisfying $\lambda_n(t) = \alpha_n$ for all $n \in \mathbf{N}$ and $t \in \mathbf{R}_+$.*
(a) *If the series $\sum_{n=1}^{\infty} 1/\alpha_n$ diverges, then the probability of explosion is equal to zero.*
(b) *If the series $\sum_{n=1}^{\infty} 1/\alpha_n$ converges, then the probability of explosion is equal to one.*

This follows from Theorems 3.4.2 and 1.2.1.

Problems

3.4.A The following are equivalent:
   (a) The claim number process is a regular Markov process and its intensities are all constant.
   (b) The claim interarrival times are independent and exponentially distributed.

3.4.B Let $\varrho : \mathbf{R}_+ \to \mathbf{R}_+$ be a continuous function which is strictly increasing and satisfies $\varrho(0) = 0$ and $\lim_{t \to \infty} \varrho(t) = \infty$. For all $t \in \mathbf{R}_+$, define

$$N_t^\varrho := N_{\varrho^{-1}(t)} \,.$$

Then $\{N_t^\varrho\}_{t \in \mathbf{R}_+}$ is a claim number process. Moreover, if $\{N_t\}_{t \in \mathbf{R}_+}$ has independent increments or is a Markov process or satisfies the Chapman-Kolmogorov equations or is regular, then the same is true for $\{N_t^\varrho\}_{t \in \mathbf{R}_+}$.

3.4.C **Operational Time:** A continuous function $\varrho : \mathbf{R}_+ \to \mathbf{R}_+$ which is strictly increasing and satisfies $\varrho(0) = 0$ and $\lim_{t \to \infty} \varrho(t) = \infty$ is an *operational time* for the claim number process $\{N_t\}_{t \in \mathbf{R}_+}$ if the claim number process $\{N_t^\varrho\}_{t \in \mathbf{R}_+}$ is homogeneous.
   Assume that the claim number process $\{N_t\}_{t \in \mathbf{R}_+}$ satisfies the Chapman–Kolmogorov equations and is regular with intensities $\{\lambda_n\}_{n \in \mathbf{N}}$, and let $\{\alpha_n\}_{n \in \mathbf{N}}$ be a sequence of real numbers in $(0, \infty)$. Then the following are equivalent:
   (a) There exists an operational time ϱ for the claim number process $\{N_t\}_{t \in \mathbf{R}_+}$ such that $\lambda_n^\varrho(t) = \alpha_n$ holds for all $n \in \mathbf{N}$ and $t \in \mathbf{R}_+$.
   (b) There exists a continuous function $\lambda : \mathbf{R}_+ \to (0, \infty)$ satisfying $\int_0^\infty \lambda(s)\,ds = \infty$ and such that $\lambda_n(t) = \alpha_n \lambda(t)$ holds for all $n \in \mathbf{N}$ and $t \in \mathbf{R}_+$.
   Hint: Use Theorem 3.2.1 and choose λ and ϱ, respectively, such that the identity $\varrho(t) = \int_0^t \lambda(s)\,ds$ holds for all $t \in \mathbf{R}_+$.

3.4.D **Operational Time:** If $\{N_t\}_{t \in \mathbf{R}_+}$ is an inhomogeneous Poisson process with intensity λ satisfying $\int_0^\infty \lambda(s)\,ds = \infty$, then there exists an operational time ϱ such that $\{N_t^\varrho\}_{t \in \mathbf{R}_+}$ is a homogeneous Poisson process with parameter 1.

3.4.E **Operational Time:** Study the explosion problem for claim number processes which are regular Markov processes and possess an operational time.

3.5   A Characterization of the Poisson Process

By Theorem 3.1.8, the Poisson process is a regular Markov process whose intensities are all identical and constant. By Theorems 3.3.2, 3.4.2, and 2.3.4, regular Markov processes whose intensities are either all identical or all constant share some of the characteristic properties of the Poisson process:
- For a regular Markov process whose intensities are all identical, the increments of the claim number process are independent and Poisson distributed.
- For a regular Markov process whose intensities are all constant, the claim interarrival times are independent and exponentially distributed.

In the first case, the intensities are related to the parameters of the distributions of the increments of the claim number process; in the second case, they are related to the parameters of the distributions of the claim interarrival times. It is therefore not surprising that the Poisson process is the only regular Markov process whose intensities are all identical and constant:

3.5.1 Theorem. *Let $\alpha \in (0, \infty)$. Then the following are equivalent:*
(a) *The claim number process $\{N_t\}_{t \in \mathbf{R}_+}$ is a regular Markov process with intensities $\{\lambda_n\}_{n \in \mathbf{N}}$ satisfying $\lambda_n(t) = \alpha$ for all $n \in \mathbf{N}$ and $t \in \mathbf{R}_+$.*
(b) *The claim number process $\{N_t\}_{t \in \mathbf{R}_+}$ has independent increments and is regular with intensities $\{\lambda_n\}_{n \in \mathbf{N}}$ satisfying $\lambda_n(t) = \alpha$ for all $n \in \mathbf{N}$ and $t \in \mathbf{R}_+$.*
(c) *The claim number process $\{N_t\}_{t \in \mathbf{R}_+}$ is a Poisson process with parameter α.*
(d) *The sequence of claim interarrival times $\{W_n\}_{n \in \mathbf{N}}$ is independent and satisfies $P_{W_n} = \mathbf{Exp}(\alpha)$ for all $n \in \mathbf{N}$.*

Proof. The equivalence of (a), (b), and (c) follows from Theorem 3.3.2, and the equivalence of (a) and (d) follows from Theorem 3.4.2. □

The equivalence of (c) and (d) in Theorem 3.5.1 is the same as the equivalence of (a) and (b) in Theorem 2.3.4, which was established by entirely different arguments; in particular, Theorem 2.3.4 was not used in the proof of Theorem 3.5.1.

Problems

3.5.A Assume that the claim number process has stationary independent increments and is regular. Then its intensities are all identical and constant.

3.5.B The following are equivalent:
(a) The claim number process is a regular Markov process and its intensities are all identical and constant.
(b) The claim number process has stationary independent increments and is regular.
(c) The claim number process is a Poisson process.
(d) The claim interarrival times are independent and identically exponentially distributed.

3.6   A Claim Number Process with Contagion

In the present section we study a regular claim number process which is homogeneous and satisfies the Chapman–Kolmogorov equations but need not be a Markov process and fails to be a Poisson process. More precisely, the increments of this claim number process fail to be independent, fail to be stationary, and fail to be Poisson distributed. In other words, this claim number process lacks each of the defining properties of the Poisson process.

3.6.1 Theorem (Positive Contagion). *Let $\alpha, \beta \in (0, \infty)$. Then the following are equivalent:*

(a) *The claim number process $\{N_t\}_{t \in \mathbf{R}_+}$ satisfies the Chapman–Kolmogorov equations and is regular with intensities $\{\lambda_n\}_{n \in \mathbf{N}}$ satisfying*

$$\lambda_n(t) = (\alpha + n - 1)\beta$$

for all $n \in \mathbf{N}$ and $t \in \mathbf{R}_+$.

(b) *The identity*

$$p_{n,n+k}(t, t+h) = \binom{\alpha + n + k - 1}{k} (e^{-\beta h})^{\alpha + n} (1 - e^{-\beta h})^k$$

holds for each admissible pair (n, t) and all $k \in \mathbf{N}_0$ and $h \in \mathbf{R}_+$.

In this case, the claim number process is homogeneous and satisfies

$$P_{N_t} = \mathbf{NB}(\alpha, e^{-\beta t})$$

for all $t \in (0, \infty)$, and the increments are neither independent nor stationary and satisfy

$$P_{N_{t+h} - N_t} = \mathbf{NB}\left(\alpha, \frac{1}{1 + e^{\beta(t+h)} - e^{\beta t}}\right)$$

for all $t \in \mathbf{R}_+$ and $h \in (0, \infty)$.

Proof. • Assume first that (a) holds.

(1) For each admissible pair (n, t) and all $h \in \mathbf{R}_+$, we have

$$\begin{aligned}
p_{n,n}(t, t+h) &= e^{-\int_t^{t+h} \lambda_{n+1}(u)\, du} \\
&= e^{-\int_t^{t+h} (\alpha+n)\beta\, du} \\
&= (e^{-\beta h})^{\alpha+n}.
\end{aligned}$$

(2) Assume now that the identity

$$p_{n,n+k}(t, t+h) = \binom{\alpha + n + k - 1}{k} (e^{-\beta h})^{\alpha + n} (1 - e^{-\beta h})^k$$

holds for some $k \in \mathbf{N}_0$ and for each admissible pair (n, t) and all $h \in \mathbf{R}_+$ (which because of (1) is the case for $k = 0$). Then we have

$$\begin{aligned}
p_{n,n+k+1}(t, t+h) &= \int_t^{t+h} p_{n,n+k}(t, u)\, \lambda_{n+k+1}(u)\, p_{n+k+1,n+k+1}(u, t+h)\, du \\
&= \int_t^{t+h} \left(\binom{\alpha + n + k - 1}{k} (e^{-\beta(u-t)})^{\alpha+n} (1 - e^{-\beta(u-t)})^k \right. \\
&\qquad\qquad \left. \cdot (\alpha + n + k)\beta \left(e^{-\beta(t+h-u)}\right)^{\alpha+n+k+1} \right) du
\end{aligned}$$

$$= \binom{\alpha + n + (k+1) - 1}{k+1} \left(e^{-\beta h}\right)^{\alpha+n}$$

$$\cdot \int_t^{t+h} (k+1)\left(1 - e^{-\beta(u-t)}\right)^k \left(e^{-\beta(t+h-u)}\right)^{k+1} \beta \, du$$

$$= \binom{\alpha + n + (k+1) - 1}{k+1} \left(e^{-\beta h}\right)^{\alpha+n}$$

$$\cdot \int_t^{t+h} (k+1)\left(e^{-\beta(t+h-u)} - e^{-\beta h}\right)^k e^{-\beta(t+h-u)} \beta \, du$$

$$= \binom{\alpha + n + (k+1) - 1}{k+1} \left(e^{-\beta h}\right)^{\alpha+n} \left(1 - e^{-\beta h}\right)^{k+1}$$

for each admissible pair (n, t) and all $h \in \mathbf{R}_+$.

(3) Because of (1) and (2), (a) implies (b).

• Assume now that (b) holds.

(1) To verify the Chapman–Kolmogorov equations, consider $(k, n, r, t) \in \mathcal{A}$ and $s \in [r, t]$ such that $P[\{N_r = k\}] > 0$.

In the case $r = t$, there is nothing to prove.

In the case $r < t$, we have

$$\sum_{m=k}^n p_{k,m}(r,s) \, p_{m,n}(s,t)$$

$$= \sum_{m=k}^n \left(\binom{\alpha + m - 1}{m - k} \left(e^{-\beta(s-r)}\right)^{\alpha+k} \left(1 - e^{-\beta(s-r)}\right)^{m-k} \right.$$

$$\left. \cdot \binom{\alpha + n - 1}{n - m} \left(e^{-\beta(t-s)}\right)^{\alpha+m} \left(1 - e^{-\beta(t-s)}\right)^{n-m} \right)$$

$$= \sum_{m=k}^n \left(\binom{\alpha + n - 1}{n - k} \left(e^{-\beta(t-r)}\right)^{\alpha+k} \left(1 - e^{-\beta(t-r)}\right)^{n-k} \right.$$

$$\left. \cdot \binom{n - k}{m - k} \left(\frac{e^{-\beta(t-s)} - e^{-\beta(t-r)}}{1 - e^{-\beta(t-r)}}\right)^{m-k} \left(\frac{1 - e^{-\beta(t-s)}}{1 - e^{-\beta(t-r)}}\right)^{(n-k)-(m-k)} \right)$$

$$= \binom{\alpha + n - 1}{n - k} \left(e^{-\beta(t-r)}\right)^{\alpha+k} \left(1 - e^{-\beta(t-r)}\right)^{n-k}$$

$$\cdot \sum_{m=k}^n \binom{n - k}{m - k} \left(\frac{e^{-\beta(t-s)} - e^{-\beta(t-r)}}{1 - e^{-\beta(t-r)}}\right)^{m-k} \left(\frac{1 - e^{-\beta(t-s)}}{1 - e^{-\beta(t-r)}}\right)^{(n-k)-(m-k)}$$

$$= \binom{\alpha + k + (n-k) - 1}{n - k} \left(e^{-\beta(t-r)}\right)^{\alpha+k} \left(1 - e^{-\beta(t-r)}\right)^{n-k}$$

$$= p_{k,n}(r,t) \, .$$

Therefore, $\{N_t\}_{t \in \mathbf{R}_+}$ satisfies the Chapman–Kolmogorov equations.

(2) To prove the assertion on regularity, consider an admissible pair (n, t).
First, since

$$
\begin{aligned}
P[\{N_t = n\}] &= p_{0,n}(0, t) \\
&= \binom{\alpha + n - 1}{n} (e^{-\beta t})^\alpha (1 - e^{-\beta t})^n ,
\end{aligned}
$$

we have $P[\{N_t = n\}] > 0$, which proves (i).
Second, since

$$
\begin{aligned}
p_{n,n}(t, t+h) &= \left(e^{-\beta h}\right)^{\alpha + n} \\
&= e^{-(\alpha + n)\beta h} ,
\end{aligned}
$$

the function $h \mapsto p_{n,n}(t, t+h)$ is continuous, which proves (ii).
Finally, we have

$$
\begin{aligned}
\lim_{h \to 0} \frac{1}{h}\left(1 - p_{n,n}(t, t+h)\right) &= \lim_{h \to 0} \frac{1}{h}\left(1 - e^{-(\alpha + n)\beta h}\right) \\
&= (\alpha + n)\beta ,
\end{aligned}
$$

as well as

$$
\begin{aligned}
\lim_{h \to 0} \frac{1}{h} p_{n,n+1}(t, t+h) &= \lim_{h \to 0} \frac{1}{h}(\alpha + n)\left(e^{-\beta h}\right)^{\alpha + n}(1 - e^{-\beta h}) \\
&= (\alpha + n)\beta .
\end{aligned}
$$

This proves (iii).
We have thus shown that $\{N_t\}_{t \in \mathbf{R}_+}$ is regular with intensities $\{\lambda_n\}_{n \in \mathbf{N}}$ satisfying
$\lambda_n(t) = (\alpha + n - 1)\beta$ for all $n \in \mathbf{N}$ and $t \in \mathbf{R}_+$.
(3) Because of (1) and (2), (b) implies (a).
• Let us now prove the final assertions.
(1) Consider $t \in \mathbf{R}_+$. For all $n \in \mathbf{N}_0$, we have

$$
\begin{aligned}
P[\{N_t = n\}] &= p_{0,n}(0, t) \\
&= \binom{\alpha + n - 1}{n} (e^{-\beta t})^\alpha (1 - e^{-\beta t})^n .
\end{aligned}
$$

This yields

$$
P_{N_t} = \mathbf{NB}\left(\alpha, e^{-\beta t}\right) .
$$

(2) Consider now $t, h \in \mathbf{R}_+$. Because of (1), we have, for all $k \in \mathbf{N}_0$,

$$
P[\{N_{t+h} - N_t = k\}]
$$

$$
= \sum_{n=0}^{\infty} P[\{N_{t+h} - N_t = k\} \cap \{N_t = n\}]
$$

$$= \sum_{n=0}^{\infty} \left(P[\{N_{t+h} = n+k\}|\{N_t = n\}] \cdot P[\{N_t = n\}] \right)$$

$$= \sum_{n=0}^{\infty} p_{n,n+k}(t, t+h) \, p_{0,n}(0, t)$$

$$= \sum_{n=0}^{\infty} \left(\binom{\alpha + n + k - 1}{k} \left(e^{-\beta h} \right)^{\alpha + n} \left(1 - e^{-\beta h} \right)^k \right.$$

$$\left. \cdot \binom{\alpha + n - 1}{n} \left(e^{-\beta t} \right)^{\alpha} \left(1 - e^{-\beta t} \right)^n \right)$$

$$= \sum_{n=0}^{\infty} \left(\binom{\alpha + k - 1}{k} \left(\frac{e^{-\beta(t+h)}}{1 - e^{-\beta h} + e^{-\beta(t+h)}} \right)^{\alpha} \left(\frac{1 - e^{-\beta h}}{1 - e^{-\beta h} + e^{-\beta(t+h)}} \right)^k \right.$$

$$\left. \cdot \binom{\alpha + k + n - 1}{n} \left(1 - e^{-\beta h} + e^{-\beta(t+h)} \right)^{\alpha + k} \left(e^{-\beta h} - e^{-\beta(t+h)} \right)^n \right)$$

$$= \binom{\alpha + k - 1}{k} \left(\frac{e^{-\beta(t+h)}}{1 - e^{-\beta h} + e^{-\beta(t+h)}} \right)^{\alpha} \left(\frac{1 - e^{-\beta h}}{1 - e^{-\beta h} + e^{-\beta(t+h)}} \right)^k$$

$$\cdot \sum_{n=0}^{\infty} \binom{\alpha + k + n - 1}{n} \left(1 - e^{-\beta h} + e^{-\beta(t+h)} \right)^{\alpha + k} \left(e^{-\beta h} - e^{-\beta(t+h)} \right)^n$$

$$= \binom{\alpha + k - 1}{k} \left(\frac{e^{-\beta(t+h)}}{1 - e^{-\beta h} + e^{-\beta(t+h)}} \right)^{\alpha} \left(\frac{1 - e^{-\beta h}}{1 - e^{-\beta h} + e^{-\beta(t+h)}} \right)^k$$

$$= \binom{\alpha + k - 1}{k} \left(\frac{1}{e^{\beta(t+h)} - e^{\beta t} + 1} \right)^{\alpha} \left(\frac{e^{\beta(t+h)} - e^{\beta t}}{e^{\beta(t+h)} - e^{\beta t} + 1} \right)^k .$$

This yields

$$P_{N_{t+h} - N_t} = \mathbf{NB}\left(\alpha, \frac{1}{1 + e^{\beta(t+h)} - e^{\beta t}} \right) .$$

(3) It is clear that $\{N_t\}_{t \in \mathbf{R}_+}$ is homogeneous.

(4) It is clear from the formula for the transition probabilities that $\{N_t\}_{t \in \mathbf{R}_+}$ cannot have independent increments. □

Comment: The previous result illustrates the fine line between Markov claim number processes and claim number processes which only satisfy the Chapman–Kolmogorov equations: By Theorem 3.4.2, the sequence of claim interarrival times $\{W_n\}_{n \in \mathbf{N}}$ is independent and satisfies $P_{W_n} = \mathbf{Exp}((\alpha + n - 1)\beta)$ for all $n \in \mathbf{N}$ if and only if the claim number process $\{N_t\}_{t \in \mathbf{R}_+}$ is a regular Markov process with intensities $\{\lambda_n\}_{n \in \mathbf{N}}$ satisfying $\lambda_n = (\alpha + n - 1)\beta$ for all $n \in \mathbf{N}$; in this case the claim number process clearly satisfies the equivalent conditions of Theorem 3.6.1.

On the other hand, the equivalent conditions of Theorem 3.6.1 involve only the two–dimensional distributions of the claim number process, which clearly cannot tell anything about whether the claim number process is a Markov process or not.

The following result justifies the term *positive contagion*:

3.6.2 Corollary (Positive Contagion). *Let $\alpha, \beta \in (0, \infty)$ and assume that the claim number process $\{N_t\}_{t \in \mathbf{R}_+}$ satisfies the Chapman–Kolmogorov equations and is regular with intensities $\{\lambda_n\}_{n \in \mathbf{N}}$ satisfying*

$$\lambda_n(t) \;=\; (\alpha + n - 1)\,\beta$$

for all $n \in \mathbf{N}$ and $t \in \mathbf{R}_+$. Then, for all $t, h \in (0, \infty)$, the function

$$n \;\longmapsto\; P[\{N_{t+h} \geq n+1\}|\{N_t = n\}]$$

is strictly increasing and independent of t.

Thus, in the situation of Corollary 3.6.2, the conditional probability of at least one claim occurring in the interval $(t, t+h]$ increases with the number of claims already occurred up to time t.

The claim number process considered in this section illustrates the importance of the negativebinomial distribution.

Another regular claim number process which is not homogeneous but has stationary increments and which is also related to the negativebinomial distribution and positive contagion will be studied in Chapter 4 below.

Problems

3.6.A   **Negative Contagion:** For $\alpha \in \mathbf{N}$, modify the definitions given in Section 3.1 by replacing the set $\mathbf{N}_0 \cup \{\infty\}$ by $\{0, 1, \ldots, \alpha\}$.
Let $\alpha \in \mathbf{N}$ and $\beta \in (0, \infty)$. Then the following are equivalent:
(a) The claim number process $\{N_t\}_{t \in \mathbf{R}_+}$ satisfies the Chapman–Kolmogorov equations and is regular with intensities $\{\lambda_n\}_{n \in \mathbf{N}}$ satisfying

$$\lambda_n(t) \;=\; (\alpha + 1 - n)\,\beta$$

for all $n \in \{1, \ldots, \alpha\}$ and $t \in \mathbf{R}_+$.
(b) The identity

$$p_{n,n+k}(t, t+h) \;=\; \binom{\alpha - n}{k}\left(1 - e^{-\beta h}\right)^k \left(e^{-\beta h}\right)^{\alpha - n - k}$$

holds for each admissible pair (n, t) and all $k \in \{0, 1, \ldots, \alpha - n\}$ and $h \in \mathbf{R}_+$.

In this case, $\{N_t\}_{t\in\mathbf{R}_+}$ is homogeneous, the increments are neither independent nor stationary, and

$$P_{N_{t+h}-N_t} = \mathbf{B}\left(\alpha, e^{-\beta t}\left(1-e^{-\beta h}\right)\right)$$

holds for all $t \in \mathbf{R}_+$ and $h \in (0,\infty)$; in particular,

$$P_{N_t} = \mathbf{B}\left(\alpha, 1-e^{-\beta t}\right)$$

holds for all $t \in (0,\infty)$.

3.6.B **Negative Contagion:** Let $\alpha \in \mathbf{N}$ and $\beta \in (0,\infty)$ and assume that the claim number process $\{N_t\}_{t\in\mathbf{R}_+}$ satisfies the Chapman–Kolmogorov equations and is regular with intensities $\{\lambda_n\}_{n\in\mathbf{N}}$ satisfying

$$\lambda_n(t) = (\alpha + 1 - n)\,\beta$$

for all $n \in \{1,\ldots,\alpha\}$ and $t \in \mathbf{R}_+$. Then, for all $t, h \in (0,\infty)$, the function

$$n \;\mapsto\; P[\{N_{t+h} \geq n+1\}|\{N_t = n\}]$$

is strictly decreasing and independent of t.

3.6.C If the claim number process has positive or negative contagion, then

$$P_{W_1} = \mathbf{Exp}(\alpha\beta)\,.$$

Compare this result with Theorem 3.4.2 and try to compute P_{W_n} for arbitrary $n \in \mathbf{N}$ or $n \in \{1,\ldots,\alpha\}$, respectively.

3.6.D If the claim number process has positive or negative contagion, change parameters by choosing $\alpha' \in (0,\infty)$ and $\beta' \in \mathbf{R}\setminus\{0\}$ such that

$$\lambda_n(t) = \alpha' + (n-1)\beta'$$

holds for all $n \in \mathbf{N}_0$ and $t \in \mathbf{R}_+$. Interpret the limiting case $\beta' = 0$.

3.6.E Extend the discussion of claim number processes with positive or negative contagion to claim number processes which satisfy the Chapman–Kolmogorov equations and are regular with intensities $\{\lambda_n\}_{n\in\mathbf{N}}$ satisfying

$$\lambda_n(t) = \lambda(t)\,(\alpha+n-1)$$

or

$$\lambda_n(t) = \lambda(t)\,(\alpha-n+1)$$

for some continuous function $\lambda : \mathbf{R}_+ \to (0,\infty)$ and $\alpha \in (0,\infty)$.
For a special case of this problem, see Problem 4.3.A.

3.7   Remarks

The relations between the different classes of claim number processes studied in this chapter are presented in the following table:

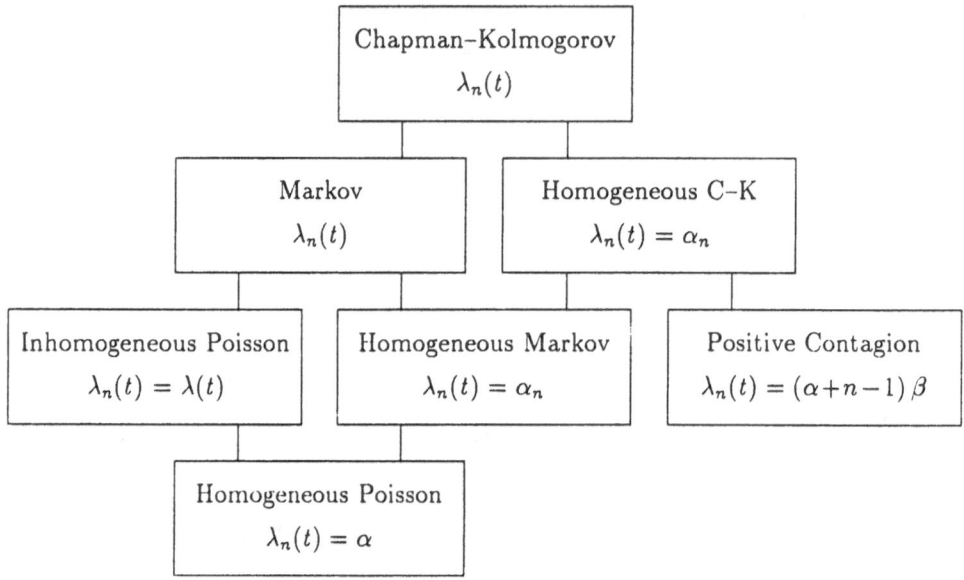

Regular Claim Number Processes

Of course, the different classes of regular claim number processes appearing in one line are not disjoint.

In Chapter 4 below, we shall study another, and rather important, claim number process which turns out to be a regular Markov process with intensities depending on time and on the number of claims already occurred; this is an example of a claim number process which is not homogeneous and has dependent stationary increments.

For a discussion of operational time, see Bühlmann [1970], Mammitzsch [1983, 1984], and Sundt [1984, 1991, 1993].

Markov claim number processes can be generalized to *semi-Markov processes* which allow to model multiple claims, to distinguish different types of claims, and to take into account claim severities; see Störmer [1970] and Nollau [1978], as well as Janssen [1977, 1982, 1984] and Janssen and DeDominicis [1984].

Chapter 4

The Mixed Claim Number Process

The choice of appropriate assumptions for the claim number process describing a portfolio of risks is a serious problem. In the present chapter we discuss a general method to reduce the problem. The basic idea is to interpret an inhomogeneous portfolio of risks as a mixture of homogeneous portfolios. The claim number process describing the inhomogeneous portfolio is then defined to be a mixture of the claim number processes describing the homogeneous portfolios such that the mixing distribution represents the structure of the inhomogeneous portfolio. We first specify the general model (Section 4.1) and then study the mixed Poisson process (Section 4.2) and, in particular, the Pólya–Lundberg process (Section 4.3).

The prerequisites required for the present chapter exceed those for the previous ones in that conditioning will be needed not only in the elementary setting but in full generality. For information on conditioning, see Bauer [1991], Billingsley [1995], and Chow and Teicher [1988].

4.1 The Model

Throughout this chapter, let $\{N_t\}_{t\in \mathbf{R}_+}$ be a claim number process and let Θ be a random variable.

Interpretation: We consider an inhomogeneous portfolio of risks. We assume that this inhomogeneous portfolio is a mixture of homogeneous portfolios of the same size which are similar but distinct, and we also assume that each of the homogeneous portfolios can be identified with a realization of the random variable Θ. This means that the distribution of Θ represents the structure of the inhomogeneous portfolio under consideration. The properties of the (unconditional) distribution of the claim number process $\{N_t\}_{t\in \mathbf{R}_+}$ are then determined by the properties of its conditional distribution with respect to Θ and by the properties of the distribution of Θ.

Accordingly, the random variable Θ and its distribution P_Θ are said to be the *structure parameter* and the *structure distribution* of the portfolio, respectively, and the claim number process $\{N_t\}_{t\in\mathbf{R}_+}$ is said to be a *mixed claim number process*.

The claim number process $\{N_t\}_{t\in\mathbf{R}_+}$ has

- *conditionally independent increments* with respect to Θ if, for all $m \in \mathbf{N}$ and $t_0, t_1, \ldots, t_m \in \mathbf{R}_+$ such that $0 = t_0 < t_1 < \ldots < t_m$, the family of increments $\{N_{t_j} - N_{t_{j-1}}\}_{j\in\{1,\ldots,m\}}$ is conditionally independent with respect to Θ, and it has
- *conditionally stationary increments* with respect to Θ if, for all $m \in \mathbf{N}$ and $t_0, t_1, \ldots, t_m, h \in \mathbf{R}_+$ such that $0 = t_0 < t_1 < \ldots < t_m$, the family of increments $\{N_{t_j+h} - N_{t_{j-1}+h}\}_{j\in\{1,\ldots,m\}}$ has the same conditional distribution with respect to Θ as $\{N_{t_j} - N_{t_{j-1}}\}_{j\in\{1,\ldots,m\}}$.

It is immediate from the definitions that a claim number process having conditionally independent increments with respect to Θ has conditionally stationary increments with respect to Θ if and only if the identity $P_{N_{t+h}-N_t|\Theta} = P_{N_h|\Theta}$ holds for all $t, h \in \mathbf{R}_+$.

4.1.1 Lemma. *If the claim number process has conditionally stationary increments with respect to Θ, then it has stationary increments.*

Proof. For all $m \in \mathbf{N}$ and $t_0, t_1, \ldots, t_m, h \in \mathbf{R}_+$ such that $0 = t_0 < t_1 < \ldots < t_m$ and for all $k_1, \ldots, k_m \in \mathbf{N}_0$, we have

$$
\begin{aligned}
P\left[\bigcap_{j=1}^m \{N_{t_j+h} - N_{t_{j-1}+h} = k_j\}\right] &= \int_\Omega P\left(\bigcap_{j=1}^m \{N_{t_j+h} - N_{t_{j-1}+h} = k_j\}\,\middle|\,\Theta(\omega)\right) dP(\omega) \\
&= \int_\Omega P\left(\bigcap_{j=1}^m \{N_{t_j} - N_{t_{j-1}} = k_j\}\,\middle|\,\Theta(\omega)\right) dP(\omega) \\
&= P\left[\bigcap_{j=1}^m \{N_{t_j} - N_{t_{j-1}} = k_j\}\right],
\end{aligned}
$$

as was to be shown. $\qquad\square$

By contrast, a claim number process having conditionally independent increments with respect to Θ need not have independent increments; see Theorem 4.2.6 below.

The following lemma is immediate from the properties of conditional expectation:

4.1.2 Lemma. *If the claim number process $\{N_t\}_{t\in\mathbf{R}_+}$ has finite expectations, then the identities*

$$E[N_t] = E[E(N_t|\Theta)]$$

and

$$\mathrm{var}\,[N_t] = E[\mathrm{var}\,(N_t|\Theta)] + \mathrm{var}\,[E(N_t|\Theta)]$$

hold for all $t \in \mathbf{R}_+$.

The second identity of Lemma 4.1.2 is called the *variance decomposition*.

4.2 The Mixed Poisson Process

The claim number process $\{N_t\}_{t\in\mathbf{R}_+}$ is a *mixed Poisson process* with parameter Θ if Θ is a random variable satisfying $P_\Theta[(0,\infty)] = 1$ and if $\{N_t\}_{t\in\mathbf{R}_+}$ has conditionally stationary independent increments with respect to Θ such that $P_{N_t|\Theta} = \mathbf{P}(t\Theta)$ holds for all $t \in (0,\infty)$.

We first collect some basic properties of the mixed Poisson process:

4.2.1 Lemma. *If the claim number process $\{N_t\}_{t\in\mathbf{R}_+}$ is a mixed Poisson process, then it has stationary increments and satisfies*

$$P[\{N_t = n\}] > 0$$

for all $t \in (0,\infty)$ and $n \in \mathbf{N}_0$.

Proof. By Lemma 4.1.1, the claim number process $\{N_t\}_{t\in\mathbf{R}_+}$ has stationary increments. Moreover, we have

$$
\begin{aligned}
P[\{N_t = n\}] &= \int_\Omega e^{-t\Theta(\omega)} \frac{(t\Theta(\omega))^n}{n!}\, dP(\omega) \\
&= \int_\mathbf{R} e^{-t\vartheta} \frac{(t\vartheta)^n}{n!}\, dP_\Theta(\vartheta)\,,
\end{aligned}
$$

and thus $P[\{N_t = n\}] > 0$. $\qquad\square$

An obvious question to ask is whether or not a mixed Poisson process can have independent increments. We shall answer this question at the end of this section.

The following result is a partial generalization of Lemma 2.3.1:

4.2.2 Lemma (Multinomial Criterion). *If the claim number process $\{N_t\}_{t\in\mathbf{R}_+}$ is a mixed Poisson process, then the identity*

$$
P\left[\bigcap_{j=1}^m \{N_{t_j} - N_{t_{j-1}} = k_j\}\,\middle|\,\{N_{t_m} = n\}\right] = \frac{n!}{\prod_{j=1}^m k_j!} \prod_{j=1}^m \left(\frac{t_j - t_{j-1}}{t_m}\right)^{k_j}
$$

holds for all $m \in \mathbf{N}$ and $t_0, t_1, \ldots, t_m \in \mathbf{R}_+$ such that $0 = t_0 < t_1 < \ldots < t_m$ and for all $n \in \mathbf{N}_0$ and $k_1, \ldots, k_m \in \mathbf{N}_0$ such that $\sum_{j=1}^m k_j = n$.

Proof. We have

$$
\begin{aligned}
&P\left[\bigcap_{j=1}^m \{N_{t_j} - N_{t_{j-1}} = k_j\} \cap \{N_{t_m} = n\}\right] \\
&= P\left[\bigcap_{j=1}^m \{N_{t_j} - N_{t_{j-1}} = k_j\}\right]
\end{aligned}
$$

$$
= \int_\Omega P\left(\bigcap_{j=1}^m \{N_{t_j} - N_{t_{j-1}} = k_j\} \,\middle|\, \Theta(\omega)\right) dP(\omega)
$$

$$
= \int_\Omega \prod_{j=1}^m P(\{N_{t_j} - N_{t_{j-1}} = k_j\} | \Theta(\omega))\, dP(\omega)
$$

$$
= \int_\Omega \prod_{j=1}^m e^{-(t_j - t_{j-1})\Theta(\omega)} \frac{((t_j - t_{j-1})\Theta(\omega))^{k_j}}{k_j!}\, dP(\omega)
$$

$$
= \frac{n!}{\prod_{j=1}^m k_j!} \prod_{j=1}^m \left(\frac{t_j - t_{j-1}}{t_m}\right)^{k_j} \cdot \int_\Omega e^{-t_m \Theta(\omega)} \frac{(t_m \Theta(\omega))^n}{n!}\, dP(\omega)
$$

as well as

$$
P[\{N_{t_m} = n\}] = \int_\Omega e^{-t_m \Theta(\omega)} \frac{(t_m \Theta(\omega))^n}{n!}\, dP(\omega)\,,
$$

and thus

$$
P\left[\bigcap_{j=1}^m \{N_{t_j} - N_{t_{j-1}} = k_j\} \,\middle|\, \{N_{t_m} = n\}\right]
$$

$$
= \frac{P\left[\bigcap_{j=1}^m \{N_{t_j} - N_{t_{j-1}} = k_j\} \cap \{N_{t_m} = n\}\right]}{P[\{N_{t_m} = n\}]}
$$

$$
= \frac{\dfrac{n!}{\prod_{j=1}^m k_j!} \prod_{j=1}^m \left(\dfrac{t_j - t_{j-1}}{t_m}\right)^{k_j} \cdot \displaystyle\int_\Omega e^{-t_m \Theta(\omega)} \frac{(t_m \Theta(\omega))^n}{n!}\, dP(\omega)}{\displaystyle\int_\Omega e^{-t_m \Theta(\omega)} \frac{(t_m \Theta(\omega))^n}{n!}\, dP(\omega)}
$$

$$
= \frac{n!}{\prod_{j=1}^m k_j!} \prod_{j=1}^m \left(\frac{t_j - t_{j-1}}{t_m}\right)^{k_j}\,,
$$

as was to be shown.   □

In the case $m = 2$, the multinomial criterion is called *Lundberg's binomial criterion*.

The multinomial criterion allows to check the assumption that the claim number process is a mixed Poisson process and is useful to compute the finite dimensional distributions of a mixed Poisson process.

As a first consequence of the multinomial criterion, we show that every mixed Poisson process is a Markov process:

4.2.3 Theorem. *If the claim number process is a mixed Poisson process, then it is a Markov process.*

Proof. Consider $m \in \mathbf{N}$, $t_1, \ldots, t_m, t_{m+1} \in (0, \infty)$, and $n_1, \ldots, n_m, n_{m+1} \in \mathbf{N}_0$ such that $t_1 < \ldots < t_m < t_{m+1}$ and $P[\bigcap_{j=1}^{m} \{N_{t_j} = n_j\}] > 0$. Define $t_0 := 0$ and $n_0 := 0$. Because of the multinomial criterion, we have

$$
P\left[\{N_{t_{m+1}} = n_{m+1}\} \,\bigg|\, \bigcap_{j=1}^{m} \{N_{t_j} = n_j\} \right]
$$

$$
= \frac{P\left[\bigcap_{j=1}^{m+1} \{N_{t_j} = n_j\} \right]}{P\left[\bigcap_{j=1}^{m} \{N_{t_j} = n_j\} \right]}
$$

$$
= \frac{P\left[\bigcap_{j=1}^{m+1} \{N_{t_j} - N_{t_{j-1}} = n_j - n_{j-1}\} \right]}{P\left[\bigcap_{j=1}^{m} \{N_{t_j} - N_{t_{j-1}} = n_j - n_{j-1}\} \right]}
$$

$$
= \frac{P\left[\bigcap_{j=1}^{m+1} \{N_{t_j} - N_{t_{j-1}} = n_j - n_{j-1}\} \,\bigg|\, \{N_{t_{m+1}} = n_{m+1}\} \right] \cdot P[\{N_{t_{m+1}} = n_{m+1}\}]}{P\left[\bigcap_{j=1}^{m} \{N_{t_j} - N_{t_{j-1}} = n_j - n_{j-1}\} \,\bigg|\, \{N_{t_m} = n_m\} \right] \cdot P[\{N_{t_m} = n_m\}]}
$$

$$
= \frac{\dfrac{n_{m+1}!}{\prod_{j=1}^{m+1}(n_j - n_{j-1})!} \displaystyle\prod_{j=1}^{m+1} \left(\dfrac{t_j - t_{j-1}}{t_{m+1}} \right)^{n_j - n_{j-1}} \cdot P[\{N_{t_{m+1}} = n_{m+1}\}]}{\dfrac{n_m!}{\prod_{j=1}^{m}(n_j - n_{j-1})!} \displaystyle\prod_{j=1}^{m} \left(\dfrac{t_j - t_{j-1}}{t_m} \right)^{n_j - n_{j-1}} \cdot P[\{N_{t_m} = n_m\}]}
$$

$$
= \binom{n_{m+1}}{n_m} \left(\frac{t_m}{t_{m+1}} \right)^{n_m} \left(\frac{t_{m+1} - t_m}{t_{m+1}} \right)^{n_{m+1} - n_m} \cdot \frac{P[\{N_{t_{m+1}} = n_{m+1}\}]}{P[\{N_{t_m} = n_m\}]}
$$

as well as

$$
P[\{N_{t_{m+1}} = n_{m+1}\} | \{N_{t_m} = n_m\}]
$$

$$
= P[\{N_{t_m} = n_m\} | \{N_{t_{m+1}} = n_{m+1}\}] \cdot \frac{P[\{N_{t_{m+1}} = n_{m+1}\}]}{P[\{N_{t_m} = n_m\}]}
$$

$$
= \binom{n_{m+1}}{n_m} \left(\frac{t_m}{t_{m+1}} \right)^{n_m} \left(\frac{t_{m+1} - t_m}{t_{m+1}} \right)^{n_{m+1} - n_m} \cdot \frac{P[\{N_{t_{m+1}} = n_{m+1}\}]}{P[\{N_{t_m} = n_m\}]},
$$

and hence

$$
P\left[\{N_{t_{m+1}} = n_{m+1}\} \,\bigg|\, \bigcap_{j=1}^{m} \{N_{t_j} = n_j\} \right] = P[\{N_{t_{m+1}} = n_{m+1}\} | \{N_{t_m} = n_m\}].
$$

This proves the assertion. $\qquad\square$

4.2.4 Theorem.   *If the claim number process is a mixed Poisson process with parameter Θ such that Θ has finite moments of any order, then it is regular and satisfies*

$$p_{n,n+k}(t,t+h) \;=\; \frac{h^k}{k!} \cdot \frac{E[e^{-(t+h)\Theta}\,\Theta^{n+k}]}{E[e^{-t\Theta}\,\Theta^n]}$$

for each admissible pair (n,t) and all $k \in \mathbf{N}_0$ and $h \in (0,\infty)$ and with intensities $\{\lambda_n\}_{n\in\mathbf{N}}$ satisfying

$$\lambda_n(t) \;=\; \frac{E[e^{-t\Theta}\,\Theta^n]}{E[e^{-t\Theta}\,\Theta^{n-1}]}$$

for all $n \in \mathbf{N}$ and $t \in \mathbf{R}_+$.

Proof.   Because of the multinomial criterion, we have

$$
\begin{aligned}
p_{n,n+k}(t,t+h) \;&=\; P[\{N_{t+h}=n+k\}|\{N_t=n\}] \\[4pt]
&=\; P[\{N_t=n\}|\{N_{t+h}=n+k\}] \cdot \frac{P[\{N_{t+h}=n+k\}]}{P[\{N_t=n\}]} \\[4pt]
&=\; \binom{n+k}{n}\left(\frac{t}{t+h}\right)^n \left(\frac{h}{t+h}\right)^k \cdot \frac{\displaystyle\int_\Omega e^{(t+h)\Theta}\frac{((t+h)\Theta)^{n+k}}{(n+k)!}\,dP}{\displaystyle\int_\Omega e^{t\Theta}\frac{(t\Theta)^n}{n!}\,dP} \\[4pt]
&=\; \frac{h^k}{k!} \cdot \frac{E[e^{-(t+h)\Theta}\,\Theta^{n+k}]}{E[e^{-t\Theta}\,\Theta^n]} \;.
\end{aligned}
$$

Let us now prove the assertion on regularity.
First, Lemma 4.2.1 yields $P[\{N_t=n\}]>0$, which proves (i).
Second, since

$$p_{n,n}(t,t+h) \;=\; \frac{E[e^{-(t+h)\Theta}\,\Theta^n]}{E[e^{-t\Theta}\,\Theta^n]}\;;$$

the function $h \mapsto p_{n,n}(t,t+h)$ is continuous, which proves (ii).
Finally, we have

$$
\begin{aligned}
\lim_{h\to 0}\frac{1}{h}\left(1-p_{n,n}(t,t+h)\right) \;&=\; \lim_{h\to 0}\frac{1}{h}\left(1-\frac{E[e^{-(t+h)\Theta}\,\Theta^n]}{E[e^{-t\Theta}\,\Theta^n]}\right) \\[4pt]
&=\; \frac{E[e^{-t\Theta}\,\Theta^{n+1}]}{E[e^{-t\Theta}\,\Theta^n]}
\end{aligned}
$$

as well as

$$
\begin{aligned}
\lim_{h\to 0}\frac{1}{h}\,p_{n,n+1}(t,t+h) \;&=\; \lim_{h\to 0}\frac{1}{h}\,h\,\frac{E[e^{-(t+h)\Theta}\,\Theta^{n+1}]}{E[e^{-t\Theta}\,\Theta^n]} \\[4pt]
&=\; \frac{E[e^{-t\Theta}\,\Theta^{n+1}]}{E[e^{-t\Theta}\,\Theta^n]}\;.
\end{aligned}
$$

This proves (iii).

We have thus shown that $\{N_t\}_{t\in\mathbf{R}_+}$ is regular with intensities $\{\lambda_n\}_{n\in\mathbf{N}}$ satisfying

$$\lambda_n(t) = \frac{E[e^{-t\Theta}\,\Theta^n]}{E[e^{-t\Theta}\,\Theta^{n-1}]}$$

for all $n \in \mathbf{N}$ and $t \in \mathbf{R}_+$. □

The following result provides another possibility to check the assumption that the claim number process is a mixed Poisson process and can be used to estimate the expectation and the variance of the structure distribution of a mixed Poisson process:

4.2.5 Lemma. *If the claim number process $\{N_t\}_{t\in\mathbf{R}_+}$ is a mixed Poisson process with parameter Θ such that Θ has finite expectation, then the identities*

$$E[N_t] = t\,E[\Theta]$$

and

$$\mathrm{var}\,[N_t] = t\,E[\Theta] + t^2\,\mathrm{var}\,[\Theta]$$

hold for all $t \in \mathbf{R}_+$; in particular, the probability of explosion is equal to zero.

Proof. The identities for the moments are immediate from Lemma 4.1.2, and the final assertion follows from Corollary 2.1.5. □

Thus, if the claim number process $\{N_t\}_{t\in\mathbf{R}_+}$ is a mixed Poisson process such that the structure distribution is nondegenerate and has finite expectation, then, for all $t \in (0,\infty)$, the variance of N_t strictly exceeds the expectation of N_t; moreover, the variance of N_t is of order t^2 while the expectation of N_t is of order t.

We can now answer the question whether a mixed Poisson process can have independent increments:

4.2.6 Theorem. *If the claim number process $\{N_t\}_{t\in\mathbf{R}_+}$ is a mixed Poisson process with parameter Θ such that Θ has finite expectation, then the following are equivalent:*
(a) *The distribution of Θ is degenerate.*
(b) *The claim number process $\{N_t\}_{t\in\mathbf{R}_+}$ has independent increments.*
(c) *The claim number process $\{N_t\}_{t\in\mathbf{R}_+}$ is an inhomogeneous Poisson process.*
(d) *The claim number process $\{N_t\}_{t\in\mathbf{R}_+}$ is a (homogeneous) Poisson process.*

Proof. Obviously, (a) implies (d), (d) implies (c), and (c) implies (b).
Because of Lemma 4.2.5 and Theorem 2.3.4, (b) implies (d).
Assume now that $\{N_t\}_{t\in\mathbf{R}_+}$ is a Poisson process. Then we have

$$E[N_t] = \mathrm{var}\,[N_t]$$

for all $t \in \mathbf{R}_+$, and Lemma 4.2.5 yields $\mathrm{var}\,[\Theta] = 0$, which means that the structure distribution is degenerate. Therefore, (d) implies (a). □

Problems

4.2.A   Assume that the claim number process is a mixed Poisson process with parameter Θ such that Θ has finite moments of any order. Then it has differentiable intensities $\{\lambda_n\}_{n\in\mathbb{N}}$ satisfying

$$\frac{\lambda_n'(t)}{\lambda_n(t)} = \lambda_n(t) - \lambda_{n+1}(t)$$

for all $n \in \mathbb{N}$ and $t \in \mathbb{R}_+$.

4.2.B   Assume that the claim number process is a mixed Poisson process with parameter Θ such that Θ has finite moments of any order. Then the following are equivalent:
(a) The intensities are all identical.
(b) The intensities are all constant.
(c) The claim number process is a homogeneous Poisson process.

4.2.C   **Estimation:** Assume that the claim number process $\{N_t\}_{t\in\mathbb{R}_+}$ is a mixed Poisson process with parameter Θ such that Θ has a nondegenerate distribution and a finite second moment, and define $\alpha = E[\Theta]/\mathrm{var}\,[\Theta]$ and $\gamma = E[\Theta]^2/\mathrm{var}\,[\Theta]$. Then the inequality

$$E\left[\left(\Theta - \frac{\gamma + N_t}{\alpha + t}\right)^2\right] \leq E[(\Theta - (a + bN_t))^2]$$

holds for all $t \in \mathbb{R}_+$ and for all $a, b \in \mathbb{R}$.

4.2.D   **Operational Time:** If the claim number process is a mixed Poisson process for which an operational time exists, then there exist $\alpha, \gamma \in (0, \infty)$ such that the intensities satisfy

$$\lambda_n(t) = \frac{\gamma + n - 1}{\alpha + t}$$

for all $n \in \mathbb{N}$ and $t \in \mathbb{R}_+$.
Hint: Use Problems 4.2.A and 3.4.C.

4.2.E   **Discrete Time Model:** The claim number process $\{N_l\}_{l\in\mathbb{N}_0}$ is a *mixed binomial process* with parameter Θ if $P_\Theta[(0, 1)]] = 1$ and if $\{N_l\}_{l\in\mathbb{N}_0}$ has conditionally stationary independent increments with respect to Θ such that $P_{N_l} = \mathbf{B}(l, \Theta)$ holds for all $l \in \mathbb{N}$.
If the claim number process is a mixed binomial process, then it has stationary increments.

4.2.F   **Discrete Time Model:** If the claim number process $\{N_l\}_{l\in\mathbb{N}_0}$ is a mixed binomial process, then the identity

$$P\left[\bigcap_{j=1}^m \{N_{l_j} - N_{l_{j-1}} = k_j\}\,\middle|\,\{N_{l_m} = n\}\right] = \prod_{j=1}^m \binom{l_j - l_{j-1}}{k_j} \cdot \binom{l_m}{n}^{-1}$$

holds for all $m \in \mathbb{N}$ and $l_0, l_1, \ldots, l_m \in \mathbb{N}_0$ such that $0 = l_0 < l_1 < \ldots < l_m$ and for all $n \in \mathbb{N}_0$ and $k_1, \ldots, k_m \in \mathbb{N}_0$ such that $k_j \leq l_j - l_{j-1}$ for all $j \in \{1, \ldots, m\}$ and such that $\sum_{j=1}^m k_j = n$.

4.2.G   Discrete Time Model: If the claim number process is a mixed binomial process, then it is a Markov process.

4.2.H   Discrete Time Model: If the claim number process $\{N_l\}_{l \in \mathbf{N}_0}$ is a mixed binomial process with parameter Θ, then the identities

$$E[N_l] \;=\; l\, E[\Theta]$$

and

$$\operatorname{var}[N_l] \;=\; l\, E[\Theta - \Theta^2] + l^2 \operatorname{var}[\Theta]$$

hold for all $l \in \mathbf{N}_0$.

4.2.I   Discrete Time Model: If the claim number process $\{N_l\}_{l \in \mathbf{N}_0}$ is a mixed binomial process with parameter Θ, then the following are equivalent:
(a) The distribution of Θ is degenerate.
(b) The claim number process $\{N_l\}_{l \in \mathbf{N}_0}$ has independent increments.
(c) The claim number process $\{N_l\}_{l \in \mathbf{N}_0}$ is a binomial process.

4.2.J   Discrete Time Model: Assume that the claim number process $\{N_l\}_{l \in \mathbf{N}_0}$ is a mixed binomial process with parameter Θ such that Θ has a nondegenerate distribution, and define $\alpha = E[\Theta - \Theta^2]/\operatorname{var}[\Theta]$ and $\gamma = \alpha\, E[\Theta]$. Then the inequality

$$E\left[\left(\Theta - \frac{\gamma + N_l}{\alpha + l}\right)^2\right] \;\leq\; E[(\Theta - (a + bN_l))^2]$$

holds for all $l \in \mathbf{N}_0$ and for all $a, b \in \mathbf{R}$.

4.3   The Pólya–Lundberg Process

The claim number process $\{N_t\}_{t \in \mathbf{R}_+}$ is a *Pólya–Lundberg process* with parameters α and γ if it is a mixed Poisson process with parameter Θ such that $P_\Theta = \mathbf{Ga}(\alpha, \gamma)$.

4.3.1 Theorem.   *If the claim number process $\{N_t\}_{t \in \mathbf{R}_+}$ is a Pólya–Lundberg process with parameters α and γ, then the identity*

$$P\left[\bigcap_{j=1}^{m} \{N_{t_j} = n_j\}\right] \;=\; \frac{\Gamma(\gamma + n_m)}{\Gamma(\gamma)\prod_{j=1}^{m}(n_j - n_{j-1})!}\left(\frac{\alpha}{\alpha + t_m}\right)^{\gamma} \prod_{j=1}^{m}\left(\frac{t_j - t_{j-1}}{\alpha + t_m}\right)^{n_j - n_{j-1}}$$

holds for all $m \in \mathbf{N}$, for all $t_0, t_1, \ldots, t_m \in \mathbf{R}_+$ such that $0 = t_0 < t_1 < \ldots < t_m$, and for all $n_0, n_1, \ldots, n_m \in \mathbf{N}_0$ such that $0 = n_0 \leq n_1 \leq \ldots \leq n_m$; in particular, the claim number process $\{N_t\}_{t \in \mathbf{R}_+}$ has stationary dependent increments and satisfies

$$P_{N_t} \;=\; \mathbf{NB}\left(\gamma, \frac{\alpha}{\alpha + t}\right)$$

for all $t \in (0, \infty)$ and

$$P_{N_{t+h} - N_t | N_t} \;=\; \mathbf{NB}\left(\gamma + N_t, \frac{\alpha + t}{\alpha + t + h}\right)$$

for all $t, h \in (0, \infty)$.

Proof.   Because of Lemma 4.2.1 and Theorem 4.2.6. is it clear that the claim number process has stationary dependent increments.
Let us now prove the remaining assertions:
(1) We have

$$P[\{N_t = n\}]$$

$$= \int_\Omega P(\{N_t = n\}|\Theta(\omega))\,dP(\omega)$$

$$= \int_\Omega e^{-t\Theta(\omega)}\,\frac{(t\Theta(\omega))^n}{n!}\,dP(\omega)$$

$$= \int_{\mathbf{R}} e^{-t\vartheta}\,\frac{(t\vartheta)^n}{n!}\,dP_\Theta(\vartheta)$$

$$= \int_{\mathbf{R}} e^{-t\vartheta}\,\frac{(t\vartheta)^n}{n!}\,\frac{\alpha^\gamma}{\Gamma(\gamma)}\,e^{-\alpha\vartheta}\,\vartheta^{\gamma-1}\,\chi_{(0,\infty)}(\vartheta)\,d\lambda(\vartheta)$$

$$= \frac{\Gamma(\gamma+n)}{\Gamma(\gamma)\,n!}\left(\frac{\alpha}{\alpha+t}\right)^\gamma\left(\frac{t}{\alpha+t}\right)^n\cdot\int_{\mathbf{R}}\frac{(\alpha+t)^{\gamma+n}}{\Gamma(\gamma+n)}\,e^{-(\alpha+t)\vartheta}\,\vartheta^{\gamma+n-1}\,\chi_{(0,\infty)}(\vartheta)\,d\lambda(\vartheta)$$

$$= \binom{\gamma+n-1}{n}\left(\frac{\alpha}{\alpha+t}\right)^\gamma\left(\frac{t}{\alpha+t}\right)^n,$$

and hence

$$P_{N_t} = \mathbf{NB}\left(\gamma,\frac{\alpha}{\alpha+t}\right).$$

(2) Because of the multinomial criterion and (1), we have

$$P\left[\bigcap_{j=1}^m\{N_{t_j} = n_j\}\right]$$

$$= P\left[\bigcap_{j=1}^m\{N_{t_j} = n_j\}\Big|\{N_{t_m} = n_m\}\right]\cdot P[\{N_{t_m} = n_m\}]$$

$$= P\left[\bigcap_{j=1}^m\{N_{t_j} - N_{t_{j-1}} = n_j - n_{j-1}\}\Big|\{N_{t_m} = n_m\}\right]\cdot P[\{N_{t_m} = n_m\}]$$

$$= \frac{n_m!}{\prod_{j=1}^m(n_j-n_{j-1})!}\prod_{j=1}^m\left(\frac{t_j-t_{j-1}}{t_m}\right)^{n_j-n_{j-1}}\cdot\binom{\gamma+n_m-1}{n_m}\left(\frac{\alpha}{\alpha+t_m}\right)^\gamma\left(\frac{t_m}{\alpha+t_m}\right)^{n_m}$$

$$= \frac{\Gamma(\gamma+n_m)}{\Gamma(\gamma)\prod_{j=1}^m(n_j-n_{j-1})!}\left(\frac{\alpha}{\alpha+t_m}\right)^\gamma\prod_{j=1}^m\left(\frac{t_j-t_{j-1}}{\alpha+t_m}\right)^{n_j-n_{j-1}}.$$

(3) Because of (2) and (1), we have

$$P[\{N_{t+h} = n + k\} \cap \{N_t = n\}]$$

$$= \frac{\Gamma(\gamma + n + k)}{\Gamma(\gamma)\, n!\, k!} \left(\frac{\alpha}{\alpha + t + h}\right)^\gamma \left(\frac{t}{\alpha + t + h}\right)^n \left(\frac{h}{\alpha + t + h}\right)^k$$

and

$$P[\{N_t = n\}] = \frac{\Gamma(\gamma + n)}{\Gamma(\gamma)\, n!} \left(\frac{\alpha}{\alpha + t}\right)^\gamma \left(\frac{t}{\alpha + t}\right)^n,$$

hence

$$P[\{N_{t+h} - N_t = k\} | \{N_t = n\}]$$

$$= \frac{P[\{N_{t+h} - N_t = k\} \cap \{N_t = n\}]}{P[\{N_t = n\}]}$$

$$= \frac{P[\{N_{t+h} = n + k\} \cap \{N_t = n\}]}{P[\{N_t = n\}]}$$

$$= \frac{\dfrac{\Gamma(\gamma + n + k)}{\Gamma(\gamma)\, n!\, k!} \left(\dfrac{\alpha}{\alpha + t + h}\right)^\gamma \left(\dfrac{t}{\alpha + t + h}\right)^n \left(\dfrac{h}{\alpha + t + h}\right)^k}{\dfrac{\Gamma(\gamma + n)}{\Gamma(\gamma)\, n!} \left(\dfrac{\alpha}{\alpha + t}\right)^\gamma \left(\dfrac{t}{\alpha + t}\right)^n}$$

$$= \binom{\gamma + n + k - 1}{k} \left(\frac{\alpha + t}{\alpha + t + h}\right)^{\gamma + n} \left(\frac{h}{\alpha + t + h}\right)^k,$$

and thus

$$P_{N_{t+h} - N_t | N_t} = \mathbf{NB}\left(\gamma + N_t, \frac{\alpha + t}{\alpha + t + h}\right).$$

This completes the proof. □

By Theorem 4.3.1, the Pólya–Lundberg process is not too difficult to handle since its finite dimensional distributions are completely known although its increments are not independent.

As an immediate consequence of Theorem 4.3.1, we see that the Pólya–Lundberg process has positive contagion:

4.3.2 Corollary (Positive Contagion). *If the claim number process $\{N_t\}_{t \in \mathbf{R}_+}$ is a Pólya–Lundberg process, then, for all $t, h \in (0, \infty)$, the function*

$$n \mapsto P[\{N_{t+h} \geq n+1\} | \{N_t = n\}]$$

is strictly increasing.

Thus, for the Pólya–Lundberg process, the conditional probability of at least one claim occurring in the interval $(t, t+h]$ increases with the number of claims already occurred up to time t.

We complete the discussion of the Pólya–Lundberg process by showing that it is a regular Markov process which is not homogeneous:

4.3.3 Corollary.   *If the claim number process* $\{N_t\}_{t \in \mathbf{R}_+}$ *is a Pólya–Lundberg process with parameters* α *and* γ, *then it is a regular Markov process satisfying*

$$p_{n,n+k}(t, t+h) = \binom{\gamma + n + k - 1}{k} \left(\frac{\alpha + t}{\alpha + t + h} \right)^{\gamma + n} \left(\frac{h}{\alpha + t + h} \right)^{k}$$

for each admissible pair (n, t) *and all* $k \in \mathbf{N}_0$ *and* $h \in \mathbf{R}_+$ *and with intensities* $\{\lambda_n\}_{n \in \mathbf{N}}$ *satisfying*

$$\lambda_n(t) = \frac{\gamma + n - 1}{\alpha + t}$$

for all $n \in \mathbf{N}$ *and* $t \in \mathbf{R}_+$; *in particular, the claim number process* $\{N_t\}_{t \in \mathbf{R}_+}$ *is not homogeneous.*

Proof.   By Theorems 4.2.3 and 4.2.4, the Pólya–Lundberg process is a regular Markov process.
By Theorem 4.3.1, we have

$$\begin{aligned}
p_{n,n+k}(t, t+h) &= P[\{N_{t+h} = n+k\} | \{N_t = n\}] \\
&= P[\{N_{t+h} - N_t = k\} | \{N_t = n\}] \\
&= \binom{\gamma + n + k - 1}{k} \left(\frac{\alpha + t}{\alpha + t + h} \right)^{\gamma + n} \left(\frac{h}{\alpha + t + h} \right)^{k} .
\end{aligned}$$

This yields

$$\begin{aligned}
\lambda_{n+1}(t) &= \lim_{h \to 0} \frac{1}{h} p_{n,n+1}(t, t+h) \\
&= \lim_{h \to 0} \frac{1}{h} (\gamma + n) \left(\frac{\alpha + t}{\alpha + t + h} \right)^{\gamma + n} \frac{h}{\alpha + t + h} \\
&= \frac{\gamma + n}{\alpha + t} ,
\end{aligned}$$

and thus

$$\lambda_n(t) = \frac{\gamma + n - 1}{\alpha + t} .$$

Since the intensities are not constant, it follows from Lemma 3.4.1 that the claim number process is not homogeneous.   □

In conclusion, the Pólya–Lundberg process is a regular Markov process which is not homogeneous and has stationary dependent increments. The discussion of the Pólya–Lundberg process thus completes the investigation of regular claim number processes satisfying the Chapman–Kolmogorov equations.

Problems

4.3.A Let $\alpha, \gamma \in (0, \infty)$. Then the following are equivalent:

(a) The claim number process $\{N_t\}_{t\in\mathbf{R}_+}$ satisfies the Chapman-Kolmogorov equations and is regular with intensities $\{\lambda_n\}_{n\in\mathbf{N}}$ satisfying

$$\lambda_n(t) = \frac{\gamma + n - 1}{\alpha + t}$$

for all $n \in \mathbf{N}$ and $t \in \mathbf{R}_+$.

(b) The identity

$$p_{n,n+k}(t, t+h) = \binom{\gamma + n + k - 1}{k} \left(\frac{\alpha + t}{\alpha + t + h}\right)^{\gamma+n} \left(\frac{h}{\alpha + t + h}\right)^k$$

holds for each admissible pair (n, t) and all $k \in \mathbf{N}_0$ and $h \in \mathbf{R}_+$.

In this case, the claim number process satisfies

$$P_{N_t} = \mathbf{NB}\left(\gamma, \frac{\alpha}{\alpha + t}\right)$$

for all $t \in (0, \infty)$,

$$P_{N_{t+h}-N_t} = \mathbf{NB}\left(\gamma, \frac{\alpha}{\alpha + h}\right)$$

for all $t, h \in (0, \infty)$, and

$$P[\{N_{t+h}-N_t = k\}|\{N_t = n\}] = \binom{\gamma + n + k - 1}{k} \left(\frac{\alpha + t}{\alpha + t + h}\right)^{\gamma+n} \left(\frac{h}{\alpha + t + h}\right)^k$$

for all $t, h \in (0, \infty)$ and all $n, k \in \mathbf{N}_0$; in particular, the claim number process has dependent increments and is not homogeneous.

Compare the result with Theorem 4.3.1 and Corollary 4.3.3.

4.3.B If the claim number process $\{N_t\}_{t\in\mathbf{R}_+}$ is a Pólya-Lundberg process with parameters α and γ, then

$$P_{W_1} = \mathbf{Par}(\alpha, \gamma).$$

Try to compute P_{W_n} and P_{T_n} for arbitrary $n \in \mathbf{N}$.

4.3.C **Prediction:** If the claim number process $\{N_t\}_{t\in\mathbf{R}_+}$ is a Pólya-Lundberg process with parameters α and γ, then the inequality

$$E\left[\left(N_{t+h} - \left(N_t + \frac{\alpha}{\alpha + t} \cdot \frac{\gamma}{\alpha} + \frac{t}{\alpha + t} \cdot \frac{N_t}{t}\right) h\right)^2\right] \leq E[(N_{t+h} - Z)^2]$$

holds for all $t, h \in (0, \infty)$ and for every random variable Z satisfying $E[Z^2] < \infty$ and $\sigma(Z) \subseteq \mathcal{F}_t$.

Interpret the quotient γ/α and compare the result with Theorem 2.3.5.

4.3.D **Prediction:** If the claim number process $\{N_t\}_{t \in \mathbf{R}_+}$ is a Pólya–Lundberg process with parameters α and γ, then the identity

$$E(N_{t+h} - N_t | N_t) = \left(\frac{\alpha}{\alpha + t} \cdot \frac{\gamma}{\alpha} + \frac{t}{\alpha + t} \cdot \frac{N_t}{t} \right) h$$

holds for all $t, h \in (0, \infty)$.

Interpret the expression in brackets and compare the result with Problem 4.3.C.

4.3.E **Estimation:** If the claim number process $\{N_t\}_{t \in \mathbf{R}_+}$ is a Pólya–Lundberg process with parameters α and γ, then the identity

$$P_{\Theta | N_t} = \mathbf{Ga}(\alpha + t, \gamma + N_t)$$

and hence

$$E(\Theta | N_t) = \frac{\gamma + N_t}{\alpha + t}$$

holds for all $t \in \mathbf{R}_+$.

Compare the result with Problems 4.3.D and 4.2.C.

4.3.F Assume that $P_\Theta = \mathbf{Ga}(\alpha, \gamma, \delta)$ with $\delta \in (0, \infty)$. If the claim number process $\{N_t\}_{t \in \mathbf{R}_+}$ is a mixed Poisson process with parameter Θ, then the identity

$$P \left[\bigcap_{j=1}^m \{N_{t_j} = n_j\} \right] = \frac{n_m!}{\prod_{j=1}^m (n_j - n_{j-1})!} \cdot \prod_{j=1}^m (t_j - t_{j-1})^{n_j - n_{j-1}} \cdot e^{-\delta t_m} \left(\frac{\alpha}{\alpha + t_m} \right)^\gamma$$

$$\cdot \sum_{k=0}^{n_m} \frac{\delta^k}{k!} \binom{\gamma + n_m - k - 1}{n_m - k} \left(\frac{1}{\alpha + t_m} \right)^{n_m - k}$$

holds for all $m \in \mathbf{N}$, for all $t_0, t_1, \ldots, t_m \in \mathbf{R}_+$ such that $0 = t_0 < t_1 < \ldots < t_m$ and for all $n_0, n_1, \ldots, n_m \in \mathbf{N}_0$ such that $0 = n_0 \le n_1 \le \ldots \le n_m$; in particular, the claim number process $\{N_t\}_{t \in \mathbf{R}_+}$ has stationary dependent increments and satisfies

$$P_{N_t} = \mathbf{Del}\left(\delta t, \gamma, \frac{\alpha}{\alpha + t} \right)$$

for all $t \in (0, \infty)$.

4.3.G Assume that $P_\Theta = \mathbf{Ga}(\alpha, \gamma, \delta)$ with $\delta \in (0, \infty)$. If the claim number process $\{N_t\}_{t \in \mathbf{R}_+}$ is a mixed Poisson process with parameter Θ, then it is a regular Markov process satisfying

$$p_{n,n+k}(t, t+h) = \binom{n+k}{k} h^k e^{-\delta h} \left(\frac{\alpha + t}{\alpha + t + h} \right)^\gamma$$

$$\cdot \frac{\displaystyle\sum_{j=0}^{n+k} \frac{\delta^j}{j!} \binom{\gamma + n + k - j - 1}{n + k - j} \left(\frac{1}{\alpha + t + h} \right)^{n+k-j}}{\displaystyle\sum_{j=0}^n \frac{\delta^j}{j!} \binom{\gamma + n - j - 1}{n - j} \left(\frac{1}{\alpha + t + h} \right)^{n-j}}$$

for each admissible pair (n, t) and all $k \in \mathbf{N}_0$ and $h \in \mathbf{R}_+$ and with intensities $\{\lambda_n\}_{n \in \mathbf{N}}$ satisfying

$$\lambda_n(t) = n \cdot \frac{\displaystyle\sum_{j=0}^{n} \frac{\delta^j}{j!} \binom{\gamma + n - j - 1}{n - j} \left(\frac{1}{\alpha + t}\right)^{n-j}}{\displaystyle\sum_{j=0}^{n-1} \frac{\delta^j}{j!} \binom{\gamma + n - 1 - j - 1}{n - 1 - j} \left(\frac{1}{\alpha + t}\right)^{n-1-j}}$$

for all $n \in \mathbf{N}$ and $t \in \mathbf{R}_+$; in particular, the claim number process $\{N_t\}_{t \in \mathbf{R}_+}$ is not homogeneous.

4.3.H   Assume that $\{N'_t\}_{t \in \mathbf{R}_+}$ and $\{N''_t\}_{t \in \mathbf{R}_+}$ are independent claim number processes such that $\{N'_t\}_{t \in \mathbf{R}_+}$ is a homogeneous Poisson process with parameter δ and $\{N''_t\}_{t \in \mathbf{R}_+}$ is a Pólya–Lundberg process with parameters α and γ. For all $t \in \mathbf{R}_+$, define

$$N_t := N'_t + N''_t .$$

Show that $\{N_t\}_{t \in \mathbf{R}_+}$ is a claim number process and compute its finite dimensional distributions.

4.3.I   **Discrete Time Model:** Assume that $P_\Theta = \mathbf{Be}(\alpha, \beta)$. If the claim number process $\{N_l\}_{l \in \mathbf{N}_0}$ is a mixed binomial process with parameter Θ, then the identity

$$P\left[\bigcap_{j=1}^{m} \{N_{l_j} = n_j\}\right] = \frac{\displaystyle\prod_{j=1}^{m} \binom{l_j - l_{j-1}}{n_j - n_{j-1}}}{\binom{l_m}{n_m}} \cdot \frac{\binom{\alpha + n_m - 1}{n_m} \binom{\beta + l_m - n_m - 1}{l_m - n_m}}{\binom{\alpha + \beta + l_m - 1}{l_m}}$$

holds for all $m \in \mathbf{N}$, for all $l_0, l_1, \ldots, l_m \in \mathbf{N}_0$ such that $0 = l_0 < l_1 < \ldots < l_m$, and for all $n_0, n_1, \ldots, n_m \in \mathbf{N}_0$ such that $0 = n_0 \leq n_1 \leq \ldots \leq n_m$ and such that $n_j \leq l_j$ holds for all $j \in \{1, \ldots, m\}$; in particular, the claim number process $\{N_l\}_{l \in \mathbf{N}_0}$ has stationary dependent increments and satisfies

$$P_{N_m} = \mathbf{NH}(m, \alpha, \beta)$$

for all $m \in \mathbf{N}$, and

$$P_{N_{m+l} - N_m \mid N_m} = \mathbf{NH}(l, \alpha + N_m, \beta + m - N_m)$$

for all $m \in \mathbf{N}_0$ and $l \in \mathbf{N}$.

4.3.J   **Discrete Time Model:** Assume that $P_\Theta = \mathbf{Be}(\alpha, \beta)$. If the claim number process $\{N_l\}_{l \in \mathbf{N}_0}$ is a mixed binomial process with parameter Θ, then the identity

$$P_{W_1}[\{k\}] = \frac{B(\alpha + 1, \beta + k - 1)}{B(\alpha, \beta)}$$

holds for all $k \in \mathbf{N}$.

Try to compute P_{W_n} and P_{T_n} for arbitrary $n \in \mathbf{N}$.

4.3.K   **Discrete Time Model:** Assume that $P_\Theta = \mathbf{Be}(\alpha, \beta)$. If the claim number process $\{N_l\}_{l\in\mathbf{N}_0}$ is a mixed binomial process with parameter Θ, then it is a Markov process satisfying

$$p_{n,n+k}(m, m+l) = \frac{\dbinom{\alpha + n + k - 1}{k}\dbinom{\beta + m - n + l - k - 1}{l - k}}{\dbinom{\alpha + \beta + m + l - 1}{l}}$$

for all $m, l \in \mathbf{N}_0$, $n \in \{0, 1, \ldots, m\}$ and $k \in \{0, 1, \ldots, m + l - n\}$; in particular, the claim number process $\{N_l\}_{l\in\mathbf{N}_0}$ is not homogeneous.

4.3.L   **Discrete Time Model:** Assume that $P_\Theta = \mathbf{Be}(\alpha, \beta)$. If the claim number process $\{N_l\}_{l\in\mathbf{N}_0}$ is a mixed binomial process with parameter Θ, then the inequality

$$E\left[\left(N_{m+l} - \left(N_m + \frac{\alpha+\beta}{\alpha+\beta+m}\cdot E[\Theta] + \frac{m}{\alpha+\beta+m}\cdot\frac{N_m}{m}\right)l\right)^2\right] \leq E[(N_{m+l} - Z)^2]$$

holds for all $m, l \in \mathbf{N}$ and for every random variable Z satisfying $E[Z^2] < \infty$ and $\sigma(Z) \subseteq \mathcal{F}_m$.
Compare the result with Problem 2.3.F.

4.3.M   **Discrete Time Model:** Assume that $P_\Theta = \mathbf{Be}(\alpha, \beta)$. If the claim number process $\{N_l\}_{l\in\mathbf{N}_0}$ is a mixed binomial process with parameter Θ, then the identity

$$E(N_{m+l} - N_m | N_m) = \left(\frac{\alpha+\beta}{\alpha+\beta+m}\cdot E[\Theta] + \frac{m}{\alpha+\beta+m}\cdot\frac{N_m}{m}\right)l$$

holds for all $m, l \in \mathbf{N}$.

4.3.N   **Discrete Time Model:** Assume that $P_\Theta = \mathbf{Be}(\alpha, \beta)$. If the claim number process $\{N_l\}_{l\in\mathbf{N}_0}$ is a mixed binomial process with parameter Θ, then the identity

$$P_{\Theta|N_m} = \mathbf{Be}(\alpha + N_m, \beta + m - N_m)$$

and hence

$$E(\Theta|N_m) = \frac{\alpha + N_m}{\alpha + \beta + m}$$

holds for all $m \in \mathbf{N}_0$.

4.4   Remarks

The background of the model discussed in this chapter may become clearer if we agree to distinguish between insurer's portfolios and abstract portfolios. An *insurer's portfolio* is, of course, a group of risks insured by the same insurance company; by contrast, an *abstract portfolio* is a group of risks which are distributed among one or more insurance companies. Homogeneous insurer's portfolios tend to be small, but homogeneous abstract portfolios may be large. Therefore, it seems to be reasonable to combine information from all insurance companies engaged in the same insurance business in order to model the claim number processes of homogeneous abstract portfolios. This gives reliable information on the conditional distribution of the claim number process of each insurer's portfolio. The single insurance company is then left with the appropriate choice of the structure distribution of its own portfolio.

The interpretation of the mixed claim number process may be extended as follows: Until now, we have assumed that an inhomogeneous insurer's portfolio is a mixture of abstract portfolios which are homogeneous. In some classes of nonlife insurance like industrial fire risks insurance, however, it is difficult to imagine portfolios which are homogeneous and sufficiently large to provide reliable statistical information. It is therefore convenient to modify the model by assuming that a rather inhomogeneous insurer's portfolio is a mixture of rather homogeneous abstract portfolios – the mathematics of mixing does not change at all. Once this generalization is accepted, we may also admit more than two levels of inhomogeneity and mix rather homogeneous portfolios to describe more and more inhomogeneous portfolios. In any case, the level of inhomogeneity of a portfolio is reflected by the variance of the structure distribution, which is equal to zero if and only if the portfolio is homogeneous.

There is still another variant of the interpretations given so far: The mixed claim number process may interpreted as the claim number process of a single risk selected at random from an inhomogeneous portfolio of risks which are similar but distinct and which can be characterized by a realization of the structure parameter which is not observable. This interpretation provides a link between claim number processes or aggregate claims processes and *experience rating* – a theory of premium calculation which, at its core, is concerned with optimal prediction of future claim numbers or claim severities of single risks, given individual claims experience in the past as well as complete or partial information on the structure of the portfolio from which the risk was selected. An example of an optimal prediction formula for future claim numbers is given in Problem 4.3.C. For an introduction to experience rating, see Sundt [1984, 1991, 1993] and Schmidt [1992].

According to Seal [1983], the history of the mixed Poisson process dates back to a paper of Dubourdieu [1938] who proposed it as a model in automobile insurance but did not compare the model with statistical data. Two years later, the mixed Poisson process became a central topic in the famous book by Lundberg [1940] who developed its mathematical theory and studied its application to sickness and accident insurance. Still another application was suggested by Hofmann [1955] who studied the mixed Poisson process as a model for workmen's compensation insurance.

For further details on the mixed Poisson process, it is worth reading Lundberg [1940]; see also Albrecht [1981], the surveys by Albrecht [1985] and Pfeifer [1986], and the recent manuscript by Grandell [1995]. Pfeifer [1982a, 1982b] and Gerber [1983] studied asymptotic properties of the claim arrival process induced by a Pólya–Lundberg process, and Pfeifer and Heller [1987] and Pfeifer [1987] characterized the mixed Poisson process in terms of the martingale property of certain transforms of the claim arrival process. The mixed Poisson process with a structure parameter having a three–parameter Gamma distribution was first considered by Delaporte [1960, 1965], and further structure distributions were discussed by Tröblinger [1961], Kupper [1962], Albrecht [1981], and Gerber [1991]. These authors, however, studied only the one–dimensional distributions of the resulting claim number processes.

In order to select a claim number process as a model for given data, it is useful to recall some criteria which are fulfilled for certain claim number processes but fail for others. The following criteria refer to the inhomogeneous Poisson process and the mixed Poisson process, each of them including the homogeneous Poisson process as a special case:

- *Independent increments*: The inhomogeneous Poisson process has independent increments, but the mixed Poisson process with a nondegenerate structure distribution has not.
- *Stationary increments*: The mixed Poisson process has stationary increments, but the inhomogeneous Poisson process with a nonconstant intensity has not.
- *Multinomial criterion*: The multinomial criterion with probabilities proportional to time intervals holds for the mixed Poisson process, but it fails for the inhomogeneous Poisson process with a nonconstant intensity.
- *Moment inequality*: The moment inequality,

$$E[N_t] \ \leq \ \text{var}\,[N_t]$$

for all $t \in (0, \infty)$, is a strict inequality for the mixed Poisson process with a nondegenerate structure distribution, but it is an equality for the inhomogeneous Poisson process.

Once the type of the claim number process has been selected according to the previous criteria, the next step should be to choose parameters and to examine the goodness of fit of the theoretical finite dimensional distributions to the empirical ones.

Automobile insurance was not only the godfather of the mixed Poisson process when it was introduced in risk theory by Dubourdieu [1938] without reference to statistical data; it still is the most important class of insurance in which the mixed Poisson process seems to be a good model for the empirical claim number process. This is indicated in the papers by Thyrion [1960], Delaporte [1960, 1965], Tröblinger [1961], Derron [1962], Bichsel [1964], and Ruohonen [1988], and in the book by Lemaire [1985]. As a rule, however, these authors considered data from a single period only and hence compared one–dimensional theoretical distributions with empirical ones. In order to model the development of claim numbers in time, it would be necessary to compare the finite dimensional distributions of selected claim number processes with those of empirical processes.

Chapter 5

The Aggregate Claims Process

In the present chapter we introduce and study the aggregate claims process. We first extend the model considered so far (Section 5.1) and then establish some general results on compound distributions (Section 5.2). It turns out that aggregate claims distributions can be determined explicitly only in a few exceptional cases. However, the most important claim number distributions can be characterized by a simple recursion formula (Section 5.3) and admit the recursive computation of aggregate claims distributions and their moments when the claim size distribution is discrete (Section 5.4).

5.1 The Model

Throughout this chapter, let $\{N_t\}_{t\in\mathbf{R}_+}$ be a claim number process and let $\{T_n\}_{n\in\mathbf{N}_0}$ be the claim arrival process induced by the claim number process. We assume that the exceptional null set is empty and that the probability of explosion is equal to zero.

Furthermore, let $\{X_n\}_{n\in\mathbf{N}}$ be a sequence of random variables. For $t \in \mathbf{R}_+$, define

$$S_t := \sum_{k=1}^{N_t} X_k$$

$$= \sum_{n=0}^{\infty} \chi_{\{N_t=n\}} \sum_{k=1}^{n} X_k \,.$$

Of course, we have $S_0 = 0$.

Interpretation:
- X_n is the *claim amount* or *claim severity* or *claim size* of the nth claim.
- S_t is the *aggregate claims amount* of the claims occurring up to time t.

Accordingly, the sequence $\{X_n\}_{n\in\mathbf{N}}$ is said to be the *claim size process*, and the family $\{S_t\}_{t\in\mathbf{R}_+}$ is said to be the *aggregate claims process* induced by the claim number process and the claim size process.

For the remainder of this chapter, we assume that the sequence $\{X_n\}_{n\in\mathbb{N}}$ is i.i.d. and that the claim number process $\{N_t\}_{t\in\mathbb{R}_+}$ and the claim size process $\{X_n\}_{n\in\mathbb{N}}$ are independent.

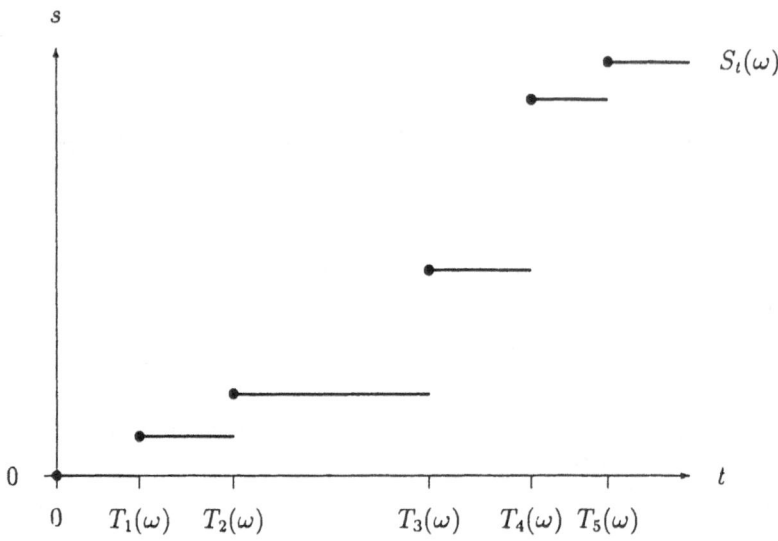

Claim Arrival Process and Aggregate Claims Process

Our first result gives a formula for the computation of aggregate claims distributions:

5.1.1 Lemma. *The identity*

$$P[\{S_t \in B\}] = \sum_{n=0}^{\infty} P[\{N_t = n\}] \, P\left[\left\{\sum_{k=1}^{n} X_k \in B\right\}\right]$$

holds for all $t \in \mathbb{R}_+$ and $B \in \mathcal{B}(\mathbb{R})$.

Proof. We have

$$
\begin{aligned}
P[\{S_t \in B\}] &= P\left[\left\{\sum_{k=1}^{N_t} X_k \in B\right\}\right] \\
&= P\left[\sum_{n=0}^{\infty} \{N_t = n\} \cap \left\{\sum_{k=1}^{n} X_k \in B\right\}\right] \\
&= \sum_{n=0}^{\infty} P[\{N_t = n\}] \, P\left[\left\{\sum_{k=1}^{n} X_k \in B\right\}\right] ,
\end{aligned}
$$

as was to be shown.   □

For $s, t \in \mathbf{R}_+$ such that $s \leq t$, the *increment* of the aggregate claims process $\{S_t\}_{t \in \mathbf{R}_+}$ on the interval $(s, t]$ is defined to be

$$S_t - S_s := \sum_{k=N_s+1}^{N_t} X_k .$$

Since $S_0 = 0$, this is in accordance the definition of S_t; in addition, we have

$$S_t(\omega) = (S_t - S_s)(\omega) + S_s(\omega) ,$$

even if $S_s(\omega)$ is infinite. For the aggregate claims process, the properties of having *independent* or *stationary* increments are defined in the same way as for the claim number process.

5.1.2 Theorem. *If the claim number process has independent increments, then the same is true for the aggregate claims process.*

Proof. Consider $m \in \mathbf{N}$, $t_0, t_1, \ldots, t_m \in \mathbf{R}_+$ such that $0 = t_0 < t_1 < \ldots < t_m$, and $B_1, \ldots, B_m \in \mathcal{B}(\mathbf{R})$. For all $n_0, n_1, \ldots, n_m \in \mathbf{N}_0$ such that $0 = n_0 \leq n_1 \leq \ldots \leq n_m$, we have

$$P\left[\left(\bigcap_{j=1}^{m}\{N_{t_j} = n_j\}\right) \cap \left(\bigcap_{j=1}^{m}\left\{\sum_{k=n_{j-1}+1}^{n_j} X_k \in B_j\right\}\right)\right]$$

$$= P\left[\bigcap_{j=1}^{m}\{N_{t_j} = n_j\}\right] \cdot P\left[\bigcap_{j=1}^{m}\left\{\sum_{k=n_{j-1}+1}^{n_j} X_k \in B_j\right\}\right]$$

$$= P\left[\bigcap_{j=1}^{m}\{N_{t_j} - N_{t_{j-1}} = n_j - n_{j-1}\}\right] \cdot P\left[\bigcap_{j=1}^{m}\left\{\sum_{k=n_{j-1}+1}^{n_j} X_k \in B_j\right\}\right]$$

$$= \prod_{j=1}^{m} P[\{N_{t_j} - N_{t_{j-1}} = n_j - n_{j-1}\}] \cdot \prod_{j=1}^{m} P\left[\left\{\sum_{k=n_{j-1}+1}^{n_j} X_k \in B_j\right\}\right]$$

$$= \prod_{j=1}^{m}\left(P[\{N_{t_j} - N_{t_{j-1}} = n_j - n_{j-1}\}] \, P\left[\left\{\sum_{k=1}^{n_j-n_{j-1}} X_k \in B_j\right\}\right]\right) .$$

This yields

$$P\left[\bigcap_{j=1}^{m}\{S_{t_j} - S_{t_{j-1}} \in B_j\}\right]$$

$$= P\left[\bigcap_{j=1}^{m}\left\{\sum_{k=N_{t_{j-1}}+1}^{N_{t_j}} X_k \in B_j\right\}\right]$$

$$= P\left[\sum_{n_1=0}^{\infty}\sum_{n_2=n_1}^{\infty}\cdots\sum_{n_m=n_{m-1}}^{\infty}\left(\bigcap_{j=1}^{m}\{N_{t_j}=n_j\}\right)\cap\left(\bigcap_{j=1}^{m}\left\{\sum_{k=N_{j-1}+1}^{N_j}X_k\in B_j\right\}\right)\right]$$

$$= \sum_{n_1=0}^{\infty}\sum_{n_2=n_1}^{\infty}\cdots\sum_{n_m=n_{m-1}}^{\infty}P\left[\left(\bigcap_{j=1}^{m}\{N_{t_j}=n_j\}\right)\cap\left(\bigcap_{j=1}^{m}\left\{\sum_{k=n_{j-1}+1}^{n_j}X_k\in B_j\right\}\right)\right]$$

$$= \sum_{n_1=0}^{\infty}\sum_{n_2=n_1}^{\infty}\cdots\sum_{n_m=n_{m-1}}^{\infty}\prod_{j=1}^{m}\left(P[\{N_{t_j}-N_{t_{j-1}}=n_j-n_{j-1}\}]\,P\left[\left\{\sum_{k=1}^{n_j-n_{j-1}}X_k\in B_j\right\}\right]\right)$$

$$= \sum_{l_1=0}^{\infty}\sum_{l_2=0}^{\infty}\cdots\sum_{l_m=0}^{\infty}\prod_{j=1}^{m}\left(P[\{N_{t_j}-N_{t_{j-1}}=l_j\}]\,P\left[\left\{\sum_{k=1}^{l_j}X_k\in B_j\right\}\right]\right)$$

$$= \prod_{j=1}^{m}\left(\sum_{l_j=0}^{\infty}P[\{N_{t_j}-N_{t_{j-1}}=l_j\}]\,P\left[\left\{\sum_{k=1}^{l_j}X_k\in B_j\right\}\right]\right).$$

The assertion follows.   □

5.1.3 Theorem.   *If the claim number process has stationary independent increments, then the same is true for the aggregate claims process.*

Proof.   By Theorem 5.1.2, the aggregate claims process has independent increments.
Consider $t, h \in \mathbf{R}_+$. For all $B \in \mathcal{B}(\mathbf{R})$, Lemma 5.1.1 yields

$$P[\{S_{t+h}-S_t\in B\}]$$

$$= P\left[\left\{\sum_{k=N_t+1}^{N_{t+h}}X_k\in B\right\}\right]$$

$$= P\left[\sum_{n=0}^{\infty}\sum_{m=0}^{\infty}\{N_t=n\}\cap\{N_{t+h}-N_t=m\}\cap\left\{\sum_{k=n+1}^{n+m}X_k\in B\right\}\right]$$

$$= \sum_{n=0}^{\infty}\sum_{m=0}^{\infty}P[\{N_t=n\}]\,P[\{N_{t+h}-N_t=m\}]\,P\left[\left\{\sum_{k=n+1}^{n+m}X_k\in B\right\}\right]$$

$$= \sum_{n=0}^{\infty}P[\{N_t=n\}]\sum_{m=0}^{\infty}P[\{N_h=m\}]\,P\left[\left\{\sum_{k=1}^{m}X_k\in B\right\}\right]$$

$$= \sum_{m=0}^{\infty}P[\{N_h=m\}]\,P\left[\left\{\sum_{k=1}^{m}X_k\in B\right\}\right]$$

$$= P[\{S_h\in B\}].$$

Therefore, the aggregate claims process has also stationary increments.   □

The assumption of Theorem 5.1.3 is fulfilled when the claim number process is a Poisson process; in this case, the aggregate claims process is also said to be a *compound Poisson process*.

Interpretation: For certain claim size distributions, the present model admits interpretations which differ from the one presented before. Of course, if the claim size distribution satisfies $P_X[\{1\}] = 1$, then the aggregate claims process is identical with the claim number process. More generally, if the claim size distribution satisfies $P_X[\mathbf{N}] = 1$, then the following interpretation is possible:
- N_t is the *number of claim events* up to time t,
- X_n is the *number of claims occurring at the nth claim event*, and
- S_t is the *total number of claims* up to time t.

This shows the possibility of applying our model to an insurance business in which two or more claims may occur simultaneously.

Further interesting interpretations are possible when the claim size distribution is a Bernoulli distribution:

5.1.4 Theorem (Thinned Claim Number Process). *If the claim size distribution is a Bernoulli distribution, then the aggregate claims process is a claim number process.*

Proof. Since P_X is a Bernoulli distribution, there exists a null set $\Omega_X \in \mathcal{F}$ such that

$$X_n(\omega) \in \{0, 1\}$$

holds for all $n \in \mathbf{N}$ and $\omega \in \Omega \setminus \Omega_X$. Furthermore, since $E[X] > 0$, the Chung–Fuchs theorem yields the existence of a null set $\Omega_\infty \in \mathcal{F}$ such that

$$\sum_{n=1}^{\infty} X_k(\omega) = \infty$$

holds for all $\omega \in \Omega \setminus \Omega_\infty$. Define

$$\Omega_S := \Omega_X \cup \Omega_\infty .$$

It now follows from the properties of the claim number process $\{N_t\}_{t \in \mathbf{R}_+}$ and the previous remarks that the aggregate claims process $\{S_t\}_{t \in \mathbf{R}_+}$ is a claim number process with exceptional null set Ω_S. □

Comment: Thinned claim number processes occur in the following situations:
- Assume that $P_X = \mathbf{B}(\eta)$. If η is interpreted as the probability for a claim to be reported, then N_t is the *number of occurred claims* and S_t is the *number of reported claims*, up to time t.

- Consider a sequence of random variables $\{Y_n\}_{n\in\mathbf{N}}$ which is i.i.d. and assume that $\{N_t\}_{t\in\mathbf{R}_+}$ and $\{Y_n\}_{n\in\mathbf{N}}$ are independent. Consider also $c \in (0,\infty)$ and assume that $\eta := P[\{Y > c\}] \in (0,1)$. Then the sequence $\{X_n\}_{n\in\mathbf{N}}$, given by $X_n := \chi_{\{Y_n>c\}}$, is i.i.d. such that $P_X = \mathbf{B}(\eta)$, and $\{N_t\}_{t\in\mathbf{R}_+}$ and $\{X_n\}_{n\in\mathbf{N}}$ are independent. If Y_n is interpreted as the size of the nth claim, then N_t is the *number of occurred claims* and S_t is the *number of large claims* (exceeding c), up to time t. This is of interest in *excess of loss reinsurance*, or *XL reinsurance* for short, where the reinsurer assumes responsibility for claims exceeding the *priority* c.

A particularly interesting result on thinned claim number processes is the following:

5.1.5 Corollary (Thinned Poisson Process). *If the claim number process is a Poisson process with parameter α and if the claim size distribution is a Bernoulli distribution with parameter η, then the aggregate claims process is a Poisson process with parameter $\alpha\eta$.*

Proof. By Theorem 5.1.3, the aggregate claims process has stationary independent increments.

Consider $t \in (0,\infty)$. By Lemma 5.1.1, we have

$$
\begin{aligned}
P[\{S_t = m\}] &= \sum_{n=0}^{\infty} P[\{N_t = n\}] \, P\left[\left\{\sum_{k=1}^{n} X_k = m\right\}\right] \\
&= \sum_{n=m}^{\infty} e^{-\alpha t} \frac{(\alpha t)^n}{n!} \binom{n}{m} \eta^m (1-\eta)^{n-m} \\
&= \sum_{n=m}^{\infty} e^{-\alpha\eta t} \frac{(\alpha\eta t)^m}{m!} e^{-\alpha(1-\eta)t} \frac{(\alpha(1-\eta)t)^{n-m}}{(n-m)!} \\
&= e^{-\alpha\eta t} \frac{(\alpha\eta t)^m}{m!} \sum_{j=0}^{\infty} e^{-\alpha(1-\eta)t} \frac{(\alpha(1-\eta)t)^j}{j!} \\
&= e^{-\alpha\eta t} \frac{(\alpha\eta t)^m}{m!}
\end{aligned}
$$

for all $m \in \mathbf{N}_0$, and hence $P_{S_t} = \mathbf{P}(\alpha\eta t)$.

Therefore, the aggregate claims process is a Poisson process with parameter $\alpha\eta$. □

We shall return to the thinned claim number process in Chapter 6 below.

Problem

5.1.A   Discrete Time Model (Thinned Binomial Process): If the claim number process is a binomial process with parameter ϑ and if the claim size distribution is a Bernoulli distribution with parameter η, then the aggregate claims process is a binomial process with parameter $\vartheta\eta$.

5.2 Compound Distributions

In this and the following sections of the present chapter we shall study the problem of computing the distribution of the aggregate claims amount S_t at a fixed time t. To this end, we simplify the notation as follows:

Let N be a random variable satisfying $P_N[\mathbf{N}_0] = 1$ and define

$$S := \sum_{k=1}^{N} X_k \,.$$

Again, the random variables N and S will be referred to as the *claim number* and the *aggregate claims amount*, respectively.

We assume throughout that N and $\{X_n\}_{n \in \mathbf{N}}$ are independent, and we maintain the assumption made before that the sequence $\{X_n\}_{n \in \mathbf{N}}$ is i. i. d. In this case, the aggregate claims distribution P_S is said to be a *compound distribution* and is denoted by

$$\mathbf{C}(P_N, P_X) \,.$$

Compound distributions are also named after the claim number distribution; for example, if P_N is a Poisson distribution, then $\mathbf{C}(P_N, P_X)$ is said to be a *compound Poisson distribution*.

The following result is a reformulation of Lemma 5.1.1:

5.2.1 Lemma. *The identity*

$$P_S[B] = \sum_{n=0}^{\infty} P_N[\{n\}] \, P_X^{*n}[B]$$

holds for all $B \in \mathcal{B}(\mathbf{R})$.

Although this formula is useful in certain special cases, it requires the computation of convolutions which may be difficult or at least time consuming. For the most important claim number distributions and for claim size distributions satisfying $P_X[\mathbf{N}_0] = 1$, recursion formulas for the aggregate claims distribution and its moments will be given in Section 5.4 below.

In some cases, it is also helpful to look at the characteristic function of the aggregate claims distribution.

5.2.2 Lemma. *The characteristic function of S satisfies*

$$\varphi_S(z) = m_N(\varphi_X(z)) \,.$$

Proof.   For all $z \in \mathbf{R}$, we have

$$
\begin{aligned}
\varphi_S(z) &= E\big[e^{izS}\big] \\
&= E\Big[e^{iz\sum_{k=1}^{N} X_k}\Big] \\
&= E\Big[\sum_{n=0}^{\infty} \chi_{\{N=n\}} e^{iz\sum_{k=1}^{n} X_k}\Big] \\
&= E\Big[\sum_{n=0}^{\infty} \chi_{\{N=n\}} \prod_{k=1}^{n} e^{izX_k}\Big] \\
&= \sum_{n=0}^{\infty} P[\{N=n\}] \prod_{k=1}^{n} E\big[e^{izX_k}\big] \\
&= \sum_{n=0}^{\infty} P[\{N=n\}]\, E\big[e^{izX}\big]^n \\
&= \sum_{n=0}^{\infty} P[\{N=n\}]\, \varphi_X(z)^n \\
&= E\big[\varphi_X(z)^N\big] \\
&= m_N(\varphi_X(z))\,,
\end{aligned}
$$

as was to be shown.   □

To illustrate the previous result, let us consider some applications.

Let us first consider the compound Poisson distribution.

5.2.3 Corollary.   *If $P_N = \mathbf{P}(\alpha)$, then the characteristic function of S satisfies*

$$
\varphi_S(z) = e^{\alpha(\varphi_X(z)-1)}\,.
$$

If the claim number distribution is either a Bernoulli distribution or a logarithmic distribution, then the computation of the compound Poisson distribution can be simplified as follows:

5.2.4 Corollary.   *For all $\alpha \in (0,\infty)$ and $\eta \in (0,1)$,*

$$
\mathbf{C}\big(\mathbf{P}(\alpha),\mathbf{B}(\eta)\big) = \mathbf{P}(\alpha\eta)\,.
$$

5.2.5 Corollary.   *For all $\alpha \in (0,\infty)$ and $\eta \in (0,1)$,*

$$
\mathbf{C}\big(\mathbf{P}(\alpha),\mathbf{Log}(\eta)\big) = \mathbf{NB}\Big(\frac{\alpha}{|\log(1-\eta)|}, 1-\eta\Big)\,.
$$

Thus, the compound Poisson distributions of Corollaries 5.2.4 and 5.2.5 are nothing else than a Poisson or a negativebinomial distribution, which are easily computed by recursion; see Theorem 5.3.1 below.

Let us now consider the *compound negativebinomial distribution.*

5.2.6 Corollary. *If $P_N = \mathbf{NB}(\alpha, \vartheta)$, then the characteristic function of S satisfies*

$$\varphi_S(z) = \left(\frac{\vartheta}{1 - (1-\vartheta)\varphi_X(z)} \right)^{\alpha} .$$

A result analogous to Corollary 5.2.4 is the following:

5.2.7 Corollary. *For all $\alpha \in (0, \infty)$ and $\vartheta, \eta \in (0, 1)$,*

$$\mathbf{C}\big(\mathbf{NB}(\alpha, \vartheta), \mathbf{B}(\eta)\big) = \mathbf{NB}\left(\alpha, \frac{\vartheta}{\vartheta + (1-\vartheta)\eta} \right) .$$

For the compound negativebinomial distribution, we have two further results:

5.2.8 Corollary. *For all $m \in \mathbf{N}$, $\vartheta \in (0, 1)$, and $\beta \in (0, \infty)$,*

$$\mathbf{C}\big(\mathbf{NB}(m, \vartheta), \mathbf{Exp}(\beta)\big) = \mathbf{C}\big(\mathbf{B}(m, 1-\vartheta), \mathbf{Exp}(\beta\vartheta)\big) .$$

5.2.9 Corollary. *For all $m \in \mathbf{N}$ and $\vartheta, \eta \in (0, 1)$,*

$$\mathbf{C}\big(\mathbf{NB}(m, \vartheta), \mathbf{Geo}(\eta)\big) = \mathbf{C}\big(\mathbf{B}(m, 1-\vartheta), \mathbf{Geo}(\eta\vartheta)\big) .$$

Corollaries 5.2.8 and 5.2.9 are of interest since the *compound binomial distribution* is determined by a finite sum.

Let us now turn to the discussion of moments of the aggregate claims distribution:

5.2.10 Lemma (Wald's Identities). *Assume that $E[N] < \infty$ and $E|X| < \infty$. Then the expectation and the variance of S exist and satisfy*

$$E[S] = E[N] \, E[X]$$

and

$$\mathrm{var}\,[S] = E[N] \, \mathrm{var}\,[X] + \mathrm{var}\,[N] \, E[X]^2 .$$

Proof. We have

$$
\begin{aligned}
E[S] &= E\left[\sum_{k=1}^{N} X_k\right] \\
&= E\left[\sum_{n=1}^{\infty} \chi_{\{N=n\}} \sum_{k=1}^{n} X_k\right] \\
&= \sum_{n=1}^{\infty} P[\{N=n\}] \, E\left[\sum_{k=1}^{n} X_k\right] \\
&= \sum_{n=1}^{\infty} P[\{N=n\}] \, n \, E[X] \\
&= E[N] \, E[X] \,,
\end{aligned}
$$

which is the first identity.
Similarly, we have

$$
\begin{aligned}
E[S^2] &= E\left[\left(\sum_{k=1}^{N} X_k\right)^2\right] \\
&= E\left[\sum_{n=1}^{\infty} \chi_{\{N=n\}} \left(\sum_{k=1}^{n} X_k\right)^2\right] \\
&= \sum_{n=1}^{\infty} P[\{N=n\}] \, E\left[\left(\sum_{k=1}^{n} X_k\right)^2\right] \\
&= \sum_{n=1}^{\infty} P[\{N=n\}] \left(\mathrm{var}\left[\sum_{k=1}^{n} X_k\right] + \left(E\left[\sum_{k=1}^{n} X_k\right]\right)^2\right) \\
&= \sum_{n=1}^{\infty} P[\{N=n\}] \left(n \, \mathrm{var}[X] + n^2 \, E[X]^2\right) \\
&= E[N] \, \mathrm{var}[X] + E[N^2] \, E[X]^2 \,,
\end{aligned}
$$

and thus

$$
\begin{aligned}
\mathrm{var}[S] &= E[S^2] - E[S]^2 \\
&= \left(E[N] \, \mathrm{var}[X] + E[N^2] \, E[X]^2\right) - \left(E[N] \, E[X]\right)^2 \\
&= E[N] \, \mathrm{var}[X] + \mathrm{var}[N] \, E[X]^2 \,,
\end{aligned}
$$

which is the second identity.   □

5.2.11 Corollary. *Assume that $P_N = \mathbf{P}(\alpha)$ and $E|X| < \infty$. Then*

$$E[S] = \alpha\, E[X]$$

and

$$\mathrm{var}\,[S] = \alpha\, E[X^2]\,.$$

The following general inequalities provide upper bounds for the *tail probabilities* $P[\{S \geq c\}]$ of the aggregate claims distribution:

5.2.12 Lemma. *Let Z be a random variable and let $h : \mathbf{R} \to \mathbf{R}_+$ be a measurable function which is strictly increasing on $\mathbf{R}_+$. Then the inequality*

$$P[\{Z \geq c\}] \leq \frac{E[h(|Z|)]}{h(c)}$$

holds for all $c \in (0, \infty)$.

Proof. By Markov's inequality, we have

$$
\begin{aligned}
P[\{Z \geq c\}] &\leq\ P[\{|Z| \geq c\}] \\
&=\ P[\{h(|Z|) \geq h(c)\}] \\
&\leq\ \frac{E[h(|Z|)]}{h(c)}\,,
\end{aligned}
$$

as was to be shown. $\qquad\square$

5.2.13 Lemma (Cantelli's Inequality). *Let Z be a random variable satisfying $E[Z^2] < \infty$. Then the inequality*

$$P[\{Z \geq E[Z] + c\}] \leq \frac{\mathrm{var}\,[Z]}{c^2 + \mathrm{var}\,[Z]}$$

holds for all $c \in (0, \infty)$.

Proof. Define $Y := Z - E[Z]$. Then we have $E[Y] = 0$ and $\mathrm{var}\,[Y] = \mathrm{var}\,[Z]$. For all $x \in (-c, \infty)$, Lemma 5.2.12 yields

$$
\begin{aligned}
P\big[\{Z \geq E[Z] + c\}\big] &=\ P[\{Y \geq c\}] \\
&=\ P\big[\{Y + x \geq c + x\}\big] \\
&\leq\ \frac{E\big[(Y + x)^2\big]}{(c + x)^2} \\
&=\ \frac{E[Y^2] + x^2}{(c + x)^2} \\
&=\ \frac{\mathrm{var}\,[Y] + x^2}{(c + x)^2} \\
&=\ \frac{\mathrm{var}\,[Z] + x^2}{(c + x)^2}\,.
\end{aligned}
$$

The last expression attains its minimum at $x = \text{var}\,[Z]/c$, and this yields

$$P[\{Z \geq E[Z] + c\}] \;\leq\; \frac{\text{var}\,[Z] + \left(\dfrac{\text{var}\,[Z]}{c}\right)^2}{\left(c + \dfrac{\text{var}\,[Z]}{c}\right)^2}$$

$$= \frac{c^2\,\text{var}\,[Z] + \left(\text{var}\,[Z]\right)^2}{\left(c^2 + \text{var}\,[Z]\right)^2}$$

$$= \frac{\text{var}\,[Z]}{c^2 + \text{var}\,[Z]}\,,$$

as was to be shown.   □

Problems

5.2.A   Let Q denote the collection of all distributions $Q : \mathcal{B}(\mathbf{R}) \to [0,1]$ satisfying $Q[\mathbf{N_0}] = 1$. For $Q, R \in Q$, define $\mathbf{C}(Q, R)$ by letting

$$\mathbf{C}(Q, R)[B] := \sum_{n=0}^{\infty} Q[\{n\}]\, R^{*n}[B]$$

for all $B \in \mathcal{B}(\mathbf{R})$. Then $\mathbf{C}(Q, R) \in Q$. Moreover, the map $\mathbf{C} : Q \times Q \to Q$ turns Q into a noncommutative semigroup with neutral element δ_1.

5.2.B   **Ammeter Transform:** For all $\alpha \in (0, \infty)$ and $\vartheta \in (0, 1)$,

$$\mathbf{C}(\mathbf{NB}(\alpha, \vartheta), P_X) = \mathbf{C}\big(\mathbf{P}(\alpha\,|\log(\vartheta)|), \mathbf{C}(\mathbf{Log}(1-\vartheta), P_X)\big)\,.$$

5.2.C   **Discrete Time Model:** For all $m \in \mathbf{N}$ and $\vartheta, \eta \in (0, 1)$,

$$\mathbf{C}\big(\mathbf{B}(m, \vartheta), \mathbf{B}(\eta)\big) = \mathbf{B}(m, \vartheta\eta)\,.$$

5.2.D   **Discrete Time Model:** For all $m \in \mathbf{N}$, $\vartheta \in (0, 1)$, and $\beta \in (0, \infty)$,

$$\mathbf{C}\big(\mathbf{B}(m, \vartheta), \mathbf{Exp}(\beta)\big) = \sum_{k=0}^{m} \binom{m}{k} \vartheta^k (1-\vartheta)^{m-k} \mathbf{Ga}(\beta, k)\,,$$

where $\mathbf{Ga}(\beta, 0) := \delta_0$. Combine the result with Corollary 5.2.8.

5.2.E   If $E[N] < \infty$ and $E|X| < \infty$, then

$$\min\{E[N], \text{var}\,[N]\} \cdot E[X^2] \;\leq\; \text{var}\,[S] \;\leq\; \max\{E[N], \text{var}\,[N]\} \cdot E[X^2]\,.$$

If $P_N = \mathbf{P}(\alpha)$, then these lower and upper bounds for $\text{var}\,[S]$ are identical.

5.2.F   If $E[N] \in (0, \infty)$, $P_X[\mathbf{R_+}] = 1$ and $E[X] \in (0, \infty)$, then

$$v^2[S] = v^2[N] + \frac{1}{E[N]}\, v^2[X]\,.$$

5.2.G   If $P_N = \mathbf{P}(\alpha)$, $P_X[\mathbf{R_+}] = 1$ and $E[X] \in (0, \infty)$, then

$$v^2[S] = \frac{1}{\alpha}\left(1 + v^2[X]\right)\,.$$

5.3   A Characterization of the Binomial, Poisson, and Negativebinomial Distributions

Throughout this section, let $Q : \mathcal{B}(\mathbf{R}) \to [0,1]$ be a distribution satisfying

$$Q[\mathbf{N}_0] = 1 .$$

For $n \in \mathbf{N}_0$, define

$$q_n := Q[\{n\}] .$$

The following result characterizes the most important claim number distributions by a simple recursion formula:

5.3.1 Theorem.   *The following are equivalent*:
(a)   *Q is either the Dirac distribution δ_0 or a binomial, Poisson, or negativebinomial distribution.*
(b)   *There exist $a, b \in \mathbf{R}$ satisfying*

$$q_n = \left(a + \frac{b}{n} \right) q_{n-1}$$

for all $n \in \mathbf{N}$.
Moreover, if Q is a binomial, Poisson, or negativebinomial distribution, then $a < 1$.

Proof.   The first part of the proof will give the following decomposition of the (a, b)–plane, which in turn will be used in the second part:

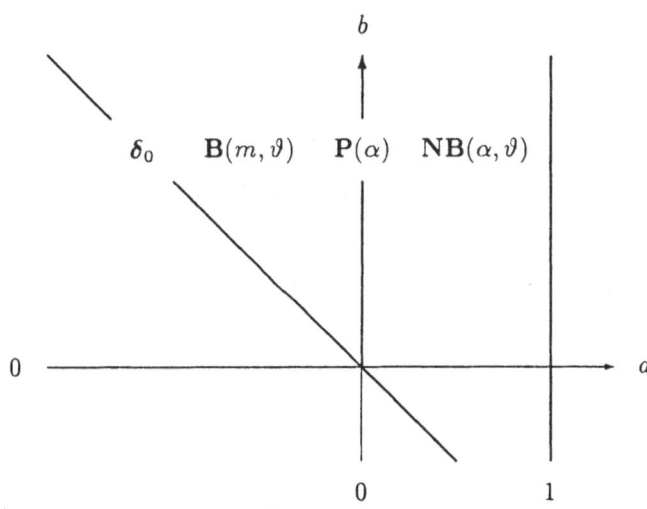

Claim Number Distributions

• Assume first that (a) holds.

If $Q = \delta_0$, then the recursion holds with $a = b = 0$.

If $Q = \mathbf{B}(m, \vartheta)$, then the recursion holds with $a = -\vartheta/(1-\vartheta)$ and $b = (m+1)\,\vartheta/(1-\vartheta)$.

If $Q = \mathbf{P}(\alpha)$, then the recursion holds with $a = 0$ and $b = \alpha$.

If $Q = \mathbf{NB}(\alpha, \vartheta)$, then the recursion holds with $a = 1-\vartheta$ and $b = (\alpha-1)(1-\vartheta)$.

Therefore, (a) implies (b).

• Assume now that (b) holds. The assumption implies that $q_0 > 0$.

Assume first that $q_0 = 1$. Then we have

$$Q = \delta_0.$$

Assume now that $q_0 < 1$. Then we have $0 < q_1 = (a + b)\, q_0$, and thus

$$a + b > 0.$$

The preceding part of the proof suggests to distinguish the three cases $a < 0$, $a = 0$, and $a > 0$.

(1) *The case $a < 0$* : Since $a + b > 0$, we have $b > 0$, and it follows that the sequence $\{a + b/n\}_{n \in \mathbf{N}}$ decreases to a. Therefore, there exists some $m \in \mathbf{N}$ satisfying $q_m > 0$ and $q_{m+1} = 0$, hence $0 = q_{m+1} = (a + b/(m+1))\, q_m$, and thus $a + b/(m+1) = 0$. This yields

$$m = \frac{a + b}{-a}$$

and $b = (m+1)(-a)$. For all $n \in \{1, \ldots, m\}$, this gives

$$
\begin{aligned}
q_n &= \left(a + \frac{b}{n} \right) q_{n-1} \\
&= \left(a + \frac{(m+1)(-a)}{n} \right) q_{n-1} \\
&= \frac{m + 1 - n}{n} (-a)\, q_{n-1} \,,
\end{aligned}
$$

and thus

$$
\begin{aligned}
q_n &= \left(\prod_{k=1}^{n} \frac{m + 1 - k}{k} \right) (-a)^n\, q_0 \\
&= \binom{m}{n} (-a)^n\, q_0 \,.
\end{aligned}
$$

Summation gives

$$
\begin{aligned}
1 &= \sum_{n=0}^{m} q_n \\
&= \sum_{n=0}^{m} \binom{m}{n} (-a)^n\, q_0 \\
&= (1 - a)^m\, q_0 \,,
\end{aligned}
$$

hence $q_0 = (1 - a)^{-m}$, and thus

$$
\begin{aligned}
q_n &= \binom{m}{n}(-a)^n q_0 \\
&= \binom{m}{n}(-a)^n (1 - a)^{-m} \\
&= \binom{m}{n}\left(\frac{-a}{1 - a}\right)^n \left(\frac{1}{1 - a}\right)^{m-n}
\end{aligned}
$$

for all $n \in \{0, \ldots, m\}$. Therefore, we have

$$
Q = \mathbf{B}\left(\frac{a + b}{-a}, \frac{-a}{1 - a}\right).
$$

(2) *The case $a = 0$* : Since $a + b > 0$, we have $b > 0$. For all $n \in \mathbf{N}$, we have

$$
q_n = \frac{b}{n} q_{n-1},
$$

and thus

$$
q_n = \frac{b^n}{n!} q_0.
$$

Summation gives

$$
\begin{aligned}
1 &= \sum_{n=0}^{\infty} q_n \\
&= \sum_{n=0}^{\infty} \frac{b^n}{n!} q_0,
\end{aligned}
$$

hence $q_0 = e^{-b}$, and thus

$$
q_n = e^{-b} \frac{b^n}{n!}
$$

for all $n \in \mathbf{N}_0$. Therefore, we have

$$
Q = \mathbf{P}(b).
$$

(3) *The case $a > 0$* : Define $c := (a+b)/a$. Then we have $c > 0$ and $b = (c-1)\,a$. For all $n \in \mathbf{N}$, this gives

$$
\begin{aligned}
q_n &= \left(a + \frac{b}{n}\right) q_{n-1} \\
&= \left(a + \frac{(c-1)a}{n}\right) q_{n-1} \\
&= \frac{c + n - 1}{n}\, a\, q_{n-1},
\end{aligned}
$$

and thus

$$q_n = \left(\prod_{k=1}^{n} \frac{c+k-1}{k}\right) a^n q_0$$

$$= \binom{c+n-1}{n} a^n q_0 \ .$$

In particular, we have $q_n \geq (1/n)ca^n q_0$. Since $\sum_{n=0}^{\infty} q_n = 1$, we must have $a < 1$. Summation yields

$$1 = \sum_{n=0}^{\infty} q_n$$

$$= \sum_{n=0}^{\infty} \binom{c+n-1}{n} a^n q_0$$

$$= (1-a)^{-c} q_0 \ ,$$

hence $q_0 = (1-a)^c$, and thus

$$q_n = \binom{c+n-1}{n} a^n q_0$$

$$= \binom{c+n-1}{n} (1-a)^c a^n \ .$$

Therefore, we have

$$Q = \mathbf{NB}\left(\frac{a+b}{a}, 1-a\right) \ .$$

We have thus shown that (b) implies (a).
• The final assertion has already been shown in the preceding part of the proof. □

Theorem 5.3.1 and its proof suggest to consider the family of all distributions Q satisfying $Q[\mathbf{N}_0] = 1$ and for which there exist $a, b \in \mathbf{R}$ satisfying $-b < a < 1$ and

$$q_n = \left(a + \frac{b}{n}\right) q_{n-1}$$

for all $n \in \mathbf{N}$ as a parametric family of distributions which consists of the binomial, Poisson, and negativebinomial distributions. Note that the Dirac distribution δ_0 is excluded since it does not determine the parameters uniquely.

Problems

5.3.A   Let Q be a nondegenerate distribution satisfying $Q[\mathbf{N_0}] = 1$ and

$$q_n = \left(a + \frac{b}{n}\right) q_{n-1}$$

for suitable $a, b \in \mathbf{R}$ and all $n \in \mathbf{N}$. Then the expectation and the variance of Q exist and satisfy

$$E[Q] = \frac{a+b}{1-a}$$

and

$$\text{var}[Q] = \frac{a+b}{(1-a)^2},$$

and hence

$$\frac{\text{var}[Q]}{E[Q]} = \frac{1}{1-a}.$$

Interpret this last identity with regard to Theorem 5.3.1 and its proof, and illustrate the result in the (a, b)–plane. Show also that

$$v^2[Q] = \frac{1}{a+b}.$$

and interpret the result.

5.3.B   If Q is a geometric or logarithmic distribution, then there exist $a, b \in \mathbf{R}$ satisfying

$$q_n = \left(a + \frac{b}{n}\right) q_{n-1}$$

for all $n \in \mathbf{N}$ such that $n \geq 2$.

5.3.C   If Q is a Delaporte distribution, then there exist $a_1, b_1, b_2 \in \mathbf{R}$ satisfying

$$q_n = \left(a_1 + \frac{b_1}{n}\right) q_{n-1} + \frac{b_2}{n} q_{n-2}$$

for all $n \in \mathbf{N}$ such that $n \geq 2$.

5.3.D   If Q is a negativehypergeometric distribution, then there exist $c_0, c_1, d_0, d_1 \in \mathbf{R}$ satisfying

$$q_n = \frac{c_0 + c_1 n}{d_0 + d_1 n} q_{n-1}$$

for all $n \in \mathbf{N}$.

5.4   The Recursions of Panjer and DePril

The aim of this section is to prove that the recursion formula for the binomial, Poisson, and negativebinomial distributions translates into recursion formulas for the aggregate claims distribution and its momemts when the claim size distribution is concentrated on $\mathbf{N_0}$.

Throughout this section, we assume that

$$P_X[\mathbf{N_0}] \;=\; 1 \,.$$

Then we have $P_S[\mathbf{N_0}] = 1$.

Comment: At the first glance, it may seem a bit confusing to assume $P_X[\mathbf{N_0}] = 1$ instead of $P_X[\mathbf{N}] = 1$. Indeed, claim severities should be strictly positive, and this is also true for the number of claims at a claim event. However, our assumption allows P_X to be an aggregate claims distribution, and this opens the possibility of applying Panjer's recursion repeatedly for a class of claim number distributions which are compound distributions; see Problem 5.4.D below.

For all $n \in \mathbf{N_0}$, define

$$
\begin{aligned}
p_n &:= P[\{N = n\}] \\
f_n &:= P[\{X = n\}] \\
g_n &:= P[\{S = n\}] \,.
\end{aligned}
$$

Then the identity of Lemma 5.2.1 can be written as

$$g_n \;=\; \sum_{k=0}^{\infty} p_k\, f_n^{*k} \,.$$

Note that the sum occurring in this formula actually extends only over a finite number of terms.

5.4.1 Lemma. *The identities*

$$f_m^{*n} \;=\; \sum_{k=0}^{m} f_{m-k}^{*(n-1)}\, f_k$$

and

$$f_m^{*n} \;=\; \frac{n}{m} \sum_{k=1}^{m} k\, f_{m-k}^{*(n-1)}\, f_k$$

hold for all $n, m \in \mathbf{N}$.

Proof. For all $j \in \{1, \ldots, n\}$ and $k \in \{0, 1, \ldots, m\}$, we have

$$
\begin{aligned}
P\!\left[\left\{\sum_{i=1}^{n} X_i = m\right\} \cap \{X_j = k\}\right]
&= P\!\left[\left\{\sum_{\substack{i=1 \\ j \neq i}}^{n} X_i = m-k\right\} \cap \{X_j = k\}\right] \\
&= P\!\left[\left\{\sum_{\substack{i=1 \\ j \neq i}}^{n} X_i = m-k\right\}\right] \cdot P[\{X_j = k\}] \\
&= f_{m-k}^{*(n-1)}\, f_k \,.
\end{aligned}
$$

This yields

$$
\begin{aligned}
f_m^{*n} &= P\left[\left\{\sum_{i=1}^{n} X_i = m\right\}\right] \\
&= \sum_{k=0}^{m} P\left[\left\{\sum_{i=1}^{n} X_i = m\right\} \cap \{X_j = k\}\right] \\
&= \sum_{k=0}^{m} f_{m-k}^{*(n-1)} f_k \,,
\end{aligned}
$$

which is the first identity, as well as

$$
\begin{aligned}
E\left[\chi_{\{\sum_{i=1}^{n} X_i = m\}} X_j\right] &= \sum_{k=0}^{m} E\left[\chi_{\{\sum_{i=1}^{n} X_i = m\} \cap \{X_j = k\}} X_j\right] \\
&= \sum_{k=1}^{m} E\left[\chi_{\{\sum_{i=1}^{n} X_i = m\} \cap \{X_j = k\}} k\right] \\
&= \sum_{k=1}^{m} k\, P\left[\left\{\sum_{i=1}^{n} X_i = m\right\} \cap \{X_j = k\}\right] \\
&= \sum_{k=1}^{m} k\, f_{m-k}^{*(n-1)} f_k
\end{aligned}
$$

for all $j \in \{1, \ldots, n\}$, and hence

$$
\begin{aligned}
f_m^{*n} &= P\left[\left\{\sum_{i=1}^{n} X_i = m\right\}\right] \\
&= E\left[\chi_{\{\sum_{i=1}^{n} X_i = m\}}\right] \\
&= E\left[\chi_{\{\sum_{i=1}^{n} X_i = m\}} \frac{1}{m} \sum_{j=1}^{n} X_j\right] \\
&= \frac{1}{m} \sum_{j=1}^{n} E\left[\chi_{\{\sum_{i=1}^{n} X_i = m\}} X_j\right] \\
&= \frac{1}{m} \sum_{j=1}^{n} \sum_{k=1}^{m} k\, f_{m-k}^{*(n-1)} f_k \\
&= \frac{n}{m} \sum_{k=1}^{m} k\, f_{m-k}^{*(n-1)} f_k \,,
\end{aligned}
$$

which is the second identity. $\qquad\square$

For the nondegenerate claim number distributions characterized by Theorem 5.3.1, we can now prove a recursion formula for the aggregate claims distribution:

5.4.2 Theorem (Panjer's Recursion). *If the distribution of N is nondegenerate and satisfies*

$$p_n = \left(a + \frac{b}{n}\right) p_{n-1}$$

for some $a, b \in \mathbf{R}$ and all $n \in \mathbf{N}$, then

$$g_0 = \begin{cases} (1 - \vartheta + \vartheta f_0)^m & \text{if} \quad P_N = \mathbf{B}(m, \vartheta) \\ e^{-\alpha(1 - f_0)} & \text{if} \quad P_N = \mathbf{P}(\alpha) \\ \left(\dfrac{\vartheta}{1 - f_0 + \vartheta f_0}\right)^\alpha & \text{if} \quad P_N = \mathbf{NB}(\alpha, \vartheta) \end{cases}$$

and the identity

$$g_n = \frac{1}{1 - a f_0} \sum_{k=1}^n \left(a + b\frac{k}{n}\right) g_{n-k} f_k$$

holds for all $n \in \mathbf{N}$; in particular, if $f_0 = 0$, then $g_0 = p_0$.

Proof.   The verification of the formula for g_0 is straightforward. For $m \in \mathbf{N}$, Lemma 5.4.1 yields

$$
\begin{aligned}
g_m &= \sum_{j=0}^\infty p_j f_m^{*j} \\
&= \sum_{j=1}^\infty p_j f_m^{*j} \\
&= \sum_{j=1}^\infty \left(a + \frac{b}{j}\right) p_{j-1} f_m^{*j} \\
&= \sum_{j=1}^\infty a\, p_{j-1} f_m^{*j} + \sum_{j=1}^\infty \frac{b}{j} p_{j-1} f_m^{*j} \\
&= \sum_{j=1}^\infty a\, p_{j-1} \sum_{k=0}^m f_{m-k}^{*(j-1)} f_k + \sum_{j=1}^\infty \frac{b}{j} p_{j-1} \frac{j}{m} \sum_{k=1}^m k\, f_{m-k}^{*(j-1)} f_k \\
&= \sum_{j=1}^\infty a\, p_{j-1} f_m^{*(j-1)} f_0 + \sum_{j=1}^\infty \sum_{k=1}^m \left(a + b\frac{k}{m}\right) p_{j-1} f_{m-k}^{*(j-1)} f_k \\
&= a f_0 \sum_{j=0}^\infty p_j f_m^{*j} + \sum_{k=1}^m \left(a + b\frac{k}{m}\right) \left(\sum_{j=0}^\infty p_j f_{m-k}^{*j}\right) f_k \\
&= a f_0\, g_m + \sum_{k=1}^m \left(a + b\frac{k}{m}\right) g_{m-k} f_k\,,
\end{aligned}
$$

and the assertion follows.   □

An analogous result holds for the moments of the aggregate claims distribution:

5.4.3 Theorem (DePril's Recursion). *If the distribution of N is nondegenerate and satisfies*

$$p_n = \left(a + \frac{b}{n}\right) p_{n-1}$$

for some $a, b \in \mathbf{R}$ and all $n \in \mathbf{N}$, then the identity

$$E[S^n] = \frac{1}{1-a} \sum_{k=1}^{n} \binom{n}{k} \left(a + b\frac{k}{n}\right) E[S^{n-k}] E[X^k]$$

holds for all $n \in \mathbf{N}$.

Proof. By Theorem 5.4.2, we have

$$
\begin{aligned}
(1 - af_0) E[S^n] &= (1 - af_0) \sum_{m=0}^{\infty} m^n g_m \\
&= \sum_{m=1}^{\infty} m^n (1 - af_0) g_m \\
&= \sum_{m=1}^{\infty} m^n \sum_{k=1}^{m} \left(a + b\frac{k}{m}\right) g_{m-k} f_k \\
&= \sum_{m=1}^{\infty} \sum_{k=1}^{m} (am^n + bkm^{n-1}) g_{m-k} f_k \\
&= \sum_{k=1}^{\infty} \sum_{m=k}^{\infty} (am^n + bkm^{n-1}) g_{m-k} f_k \\
&= \sum_{k=1}^{\infty} \sum_{l=0}^{\infty} (a(k+l)^n + bk(k+l)^{n-1}) g_l f_k \,,
\end{aligned}
$$

and hence

$$
\begin{aligned}
E[S^n] &= af_0 E[S^n] + (1 - af_0) E[S^n] \\
&= af_0 \sum_{l=0}^{\infty} l^n g_l + \sum_{k=1}^{\infty} \sum_{l=0}^{\infty} (a(k+l)^n + bk(k+l)^{n-1}) g_l f_k \\
&= \sum_{k=0}^{\infty} \sum_{l=0}^{\infty} (a(k+l)^n + bk(k+l)^{n-1}) g_l f_k \\
&= \sum_{k=0}^{\infty} \sum_{l=0}^{\infty} \left(a \sum_{j=0}^{n} \binom{n}{j} l^{n-j} k^j + b \sum_{j=0}^{n-1} \binom{n-1}{j} l^{n-1-j} k^{j+1}\right) g_l f_k
\end{aligned}
$$

$$
\begin{aligned}
&= a \sum_{j=0}^{n} \binom{n}{j} E[S^{n-j}] E[X^j] + b \sum_{j=0}^{n-1} \binom{n-1}{j} E[S^{n-1-j}] E[X^{j+1}] \\
&= a E[S^n] + a \sum_{j=1}^{n} \binom{n}{j} E[S^{n-j}] E[X^j] + b \sum_{j=1}^{n} \binom{n-1}{j-1} E[S^{n-j}] E[X^j] \\
&= a E[S^n] + \sum_{j=1}^{n} \binom{n}{j} \left(a + b \frac{j}{n} \right) E[S^{n-j}] E[X^j] \,,
\end{aligned}
$$

and the assertion follows.   □

In view of Wald's identities, DePril's recursion is of interest primarily for higher order moments of the aggregate claims distribution.

Problems

5.4.A   Use DePril's recursion to solve Problem 5.3.A.

5.4.B   Use DePril's recursion to obtain Wald's identities in the case where $P_X[N_0] = 1$.

5.4.C   Discuss the Ammeter transform in the case where $P_X[N_0] = 1$.

5.4.D   Assume that $P_X[N_0] = 1$. If P_N is nondegenerate and satisfies $P_N = \mathbf{C}(Q_1, Q_2)$, then

$$
P_S = \mathbf{C}(P_N, P_X) = \mathbf{C}(\mathbf{C}(Q_1, Q_2), P_X) = \mathbf{C}(Q_1, \mathbf{C}(Q_2, P_X)) \,.
$$

In particular, if each of Q_1 and Q_2 is a binomial, Poisson, or negativebinomial distribution, then P_S can be computed by applying Panjer's recursion twice. Extend these results to the case where P_N is obtained by repeated compounding.

5.5   Remarks

Corollary 5.2.5 is due to Quenouille [1949].

Corollary 5.2.8 is due to Panjer and Willmot [1981]; for the case $m = 1$, see also Lundberg [1940].

For an extension of Wald's identities to the case where the claim severities are not i. i. d., see Rhiel [1985].

Because of the considerable freedom in the choice of the function h, Lemma 5.2.12 is a flexible instrument to obtain upper bounds on the tail probabilities of the aggregate claims distribution. Particular bounds were obtained by Runnenburg and Goovaerts [1985] in the case where the claim number distribution is either a Poisson or a negativebinomial distribution; see also Kaas and Goovaerts [1986] for the case where the claim size is bounded. Under certain assumptions on the claim size distribution, an exponential bound on the tail probabilities of the aggregate claims distribution was obtained by Willmot and Lin [1994]; see also Gerber [1994], who proved their result by a martingale argument, and Michel [1993a], who considered the Poisson case.

The Ammeter transform is due to Ammeter [1948].

Theorem 5.3.1 is well–known; see Johnson and Kotz [1969] and Sundt and Jewell [1981].

Theorem 5.4.2 is due to Panjer [1981]; for the Poisson case, see also Shumway and Gurland [1960] and Adelson [1966]. Computational aspects of Panjer's recursion were discussed by Panjer and Willmot [1986], and numerical stability of Panjer's recursion was recently studied by Panjer and Wang [1993], who defined stability in terms of an index of error propagation and showed that the recursion is stable in the Poisson case and in the negativebinomial case but unstable in the binomial case; see also Wang and Panjer [1994].

There are two important extensions of Panjer's recursion:
- Sundt [1992] obtained a recursion for the aggregate claims distribution when the claim number distribution satisfies

$$p_n = \sum_{i=1}^{k} \left(a_i + \frac{b_i}{n} \right) p_{n-i}$$

for all $n \in \mathbf{N}$ such that $n \geq m$, where $k, m \in \mathbf{N}$, $a_i, b_i \in \mathbf{R}$, and $p_{n-i} := 0$ for all $i \in \{1, \ldots, k\}$ such that $i > n$; see also Sundt and Jewell [1981] for the case $m = 2$ and $k = 1$, and Schröter [1990] for the case $m = 1$, $k = 2$, and $a_2 = 0$. Examples of claim number distributions satisfying the above recursion are the geometric, logarithmic, and Delaporte distributions. A characterization of the claim number distributions satisfying the recursion with $m = 2$ and $k = 1$ was given by Willmot [1988].
- Hesselager [1994] obtained a recursion for the aggregate claims distribution when the claim number distribution satisfies

$$p_n = \frac{\displaystyle\sum_{j=0}^{l} c_j n^j}{\displaystyle\sum_{j=0}^{l} d_j n^j} \, p_{n-1}$$

for all $n \in \mathbf{N}$, where $l \in \mathbf{N}$ and $c_j, d_j \in \mathbf{R}$ such that $d_j \neq 0$ for some $j \in \{0, \ldots, l\}$; see also Panjer and Willmot [1982] and Willmot and Panjer [1987], who introduced this class and obtained recursions for $l = 1$ and $l = 2$, and Wang and

Sobrero [1994], who extended Hesselager's recursion to a more general class of claim number distributions. An example of a claim number distribution satisfying the above recursion is the negativehypergeometric distribution.

A common extension of the previous classes of claim number distributions is given by the class of claim number distributions satisfying

$$
p_n \;=\; \sum_{i=1}^{k} \frac{\displaystyle\sum_{j=0}^{l} c_{ij} n^j}{\displaystyle\sum_{j=0}^{l} d_{ij} n^j}\, p_{n-i}
$$

for all $n \in \mathbf{N}$ such that $n \geq m$, where $k, l, m \in \mathbf{N}$, $c_{ij}, d_{ij} \in \mathbf{R}$ such that for each $i \in \{1, \ldots, k\}$ there exists some $j \in \{0, \ldots, l\}$ satisfying $d_{ij} \neq 0$, and $p_{n-i} := 0$ for all $i \in \{1, \ldots, k\}$ such that $i > n$. A recursion for the aggregate claims distribution in this general case is not yet known; for the case $l = m = 2$, which applies to certain mixed Poisson distributions, see Willmot [1986] and Willmot and Panjer [1987].

Further important extensions of Panjer's recursion were obtained by Kling and Goovaerts [1993] and Ambagaspitiya [1995]; for special cases of their results, see Gerber [1991], Goovaerts and Kaas [1991], and Ambagaspitiya and Balakrishnan [1994].

The possibility of evaluating the aggregate claims distribution by repeated recursion when the claim number distribution is a compound distribution was first mentioned by Willmot and Sundt [1989] in the case of the Delaporte distribution; see also Michel [1993b].

Theorem 5.4.3 is due to DePril [1986]; for the Poisson case, see also Goovaerts, DeVylder, and Haezendonck [1984]. DePril actually obtained a result more general than Theorem 5.4.3, namely, a recursion for the moments of $S-c$ with $c \in \mathbf{R}$. Also, Kaas and Goovaerts [1985] obtained a recursion for the moments of the aggregate claims distribution when the claim number distribution is arbitrary.

For a discussion of further aspects concerning the computation or approximation of the aggregate claims distribution, see Hipp and Michel [1990] and Schröter [1995] and the references given there.

Let us finally remark that Scheike [1992] studied the aggregate claims process in the case where the claim arrival times and the claim severities may depend on the past through earlier claim arrival times and claim severities.

Chapter 6

The Risk Process in Reinsurance

In the present chapter we introduce the notion of a risk process (Section 6.1) and study the permanence properties of risk processes under thinning (Section 6.2), decomposition (Section 6.3), and superposition (Section 6.4). These problems are of interest in reinsurance.

6.1 The Model

A pair $(\{N_t\}_{t\in\mathbf{R}_+}, \{X_n\}_{n\in\mathbf{N}})$ is a *risk process* if
- $\{N_t\}_{t\in\mathbf{R}_+}$ is a claim number process,
- the sequence $\{X_n\}_{n\in\mathbf{N}}$ is i. i. d., and
- the pair $(\{N_t\}_{t\in\mathbf{R}_+}, \{X_n\}_{n\in\mathbf{N}})$ is independent.

In the present chapter, we shall study the following problems which are of interest in reinsurance:

First, for a risk process $(\{N_t\}_{t\in\mathbf{R}_+}, \{X_n\}_{n\in\mathbf{N}})$ and a set $C \in \mathcal{B}(\mathbf{R})$, we study the number and the (conditional) claim size distribution of claims with claim size in C. These quantities, which will be given an exact definition in Section 6.2, are of interest in excess of loss reinsurance where the reinsurer is concerned with a portfolio of large claims exceeding a priority $c \in (0, \infty)$.

Second, for a risk process $(\{N_t\}_{t\in\mathbf{R}_+}, \{X_n\}_{n\in\mathbf{N}})$ and a set $C \in \mathcal{B}(\mathbf{R})$, we study the relation between two thinned risk processes, one being generated by the claims with claim size in C and the other one being generated by the claims with claim size in the complement $\overline{C}$. This is, primarily, a mathematical problem which emerges quite naturally from the problem mentioned before.

Finally, for risk processes $(\{N_t'\}_{t\in\mathbf{R}_+}, \{X_n'\}_{n\in\mathbf{N}})$ and $(\{N_t''\}_{t\in\mathbf{R}_+}, \{X_n''\}_{n\in\mathbf{N}})$ which are independent, we study the total number and the claim size distribution of all claims which are generated by either of these risk processes. These quantities, which will be made precise in Section 6.4, are of interest to the reinsurer who forms a portfolio by combining two independent portfolios obtained from different direct insurers in order to pass from small portfolios to a larger one.

In either case, it is of interest to know whether the transformation of risk processes under consideration yields new risk processes of the same type as the original ones.

6.2   Thinning a Risk Process

Throughout this section, let $(\{N_t\}_{t\in\mathbf{R}_+}, \{X_n\}_{n\in\mathbf{N}})$ be a risk process and consider $C \in \mathcal{B}(\mathbf{R})$. Define

$$\eta \ := \ P[\{X \in C\}] \, .$$

We assume that the probability of explosion is equal to zero and that $\eta \in (0,1)$.

Let us first consider the thinned claim number process:

For all $t \in \mathbf{R}_+$, define

$$N_t' \ := \ \sum_{n=1}^{N_t} \chi_{\{X_n \in C\}} \, .$$

Thus, $\{N_t'\}_{t\in\mathbf{R}_+}$ is a particular aggregate claims process.

6.2.1 Theorem (Thinned Claim Number Process).   *The family $\{N_t'\}_{t\in\mathbf{R}_+}$ is a claim number process.*

This follows from Theorem 5.1.4.

Let us now consider the thinned claim size process, that is, the sequence of claim severities taking their values in C.

Let $\nu_0 := 0$. For all $l \in \mathbf{N}$, define

$$\nu_l \ := \ \inf\{n \in \mathbf{N} \mid \nu_{l-1} < n, \ X_n \in C\}$$

and let $\mathcal{H}(l)$ denote the collection of all sequences $\{n_j\}_{j\in\{1,\dots,l\}} \subseteq \mathbf{N}$ which are strictly increasing. For $H = \{n_j\}_{j\in\{1,\dots,l\}} \in \mathcal{H}(l)$, define $J(H) := \{1,\dots,n_l\}\setminus H$.

6.2.2 Lemma.   *The identities*

$$\bigcap_{j=1}^{l}\{\nu_j = n_j\} \ = \ \bigcap_{n\in H}\{X_n \in C\} \cap \bigcap_{n\in J(H)}\{X_n \notin C\}$$

and

$$P\left[\bigcap_{j=1}^{l}\{\nu_j = n_j\}\right] \ = \ \eta^l(1-\eta)^{n_l-l}$$

hold for all $l \in \mathbf{N}$ and for all $H = \{n_j\}_{j\in\{1,\dots,l\}} \in \mathcal{H}(l)$.

It is clear that, for each $l \in \mathbf{N}$, the family $\{\bigcap_{j=1}^{l}\{\nu_j = n_j\}\}_{H \in \mathcal{H}(l)}$ is disjoint. The following lemma shows that it is, up to a null set, even a partition of Ω:

6.2.3 Corollary. *The identity*

$$\sum_{H \in \mathcal{H}(l)} P\left[\bigcap_{j=1}^{l}\{\nu_j = n_j\}\right] = 1$$

holds for all $l \in \mathbf{N}$.

Proof. By induction, we have

$$\sum_{H \in \mathcal{H}(l)} \eta^l (1-\eta)^{n_l - l} = 1$$

for all $l \in \mathbf{N}$. The assertion now follows from Lemma 6.2.2. $\qquad \square$

The basic idea for proving the principal results of this section will be to compute the probabilities of the events of interest from their conditional probabilities with respect to events from the partition $\{\bigcap_{j=1}^{l}\{\nu_j = n_j\}\}_{H \in \mathcal{H}(l)}$ with suitable $l \in \mathbf{N}$.

By Corollary 6.2.3, each ν_n is finite. For all $n \in \mathbf{N}$, define

$$X_n' := \sum_{k=1}^{\infty} \chi_{\{\nu_n = k\}} X_k .$$

Then we have $\sigma(\{X_n'\}_{n \in \mathbf{N}}) \subseteq \sigma(\{X_n\}_{n \in \mathbf{N}})$.

The following lemma provides the technical tool for the proofs of all further results of this section:

6.2.4 Lemma. *The identity*

$$P\left[\bigcap_{i=1}^{k}\{X_i' \in B_i\} \cap \bigcap_{j=1}^{l}\{\nu_j = n_j\}\right] = \prod_{i=1}^{k} P[\{X \in B_i\}|\{X \in C\}] \cdot \eta^l (1-\eta)^{n_l - l}$$

holds for all $k, l \in \mathbf{N}$ such that $k \leq l$, for all $B_1, \ldots, B_k \in \mathcal{B}(\mathbf{R})$, and for every sequence $\{n_j\}_{j \in \{1, \ldots, l\}} \in \mathcal{H}(l)$.

Proof. For every sequence $H = \{n_j\}_{j \in \{1, \ldots, l\}} \in \mathcal{H}(l)$, we have

$$P\left[\bigcap_{i=1}^{k}\{X_i' \in B_i\} \cap \bigcap_{j=1}^{l}\{\nu_j = n_j\}\right]$$

$$= P\left[\bigcap_{i=1}^{k}\{X_{n_i} \in B_i\} \cap \bigcap_{j=1}^{l}\{\nu_j = n_j\}\right]$$

$$= P\left[\bigcap_{i=1}^{k}\{X_{n_i}\in B_i\}\cap\bigcap_{j=1}^{l}\{X_{n_j}\in C\}\cap\bigcap_{n\in J(H)}\{X_n\notin C\}\right]$$

$$= P\left[\bigcap_{i=1}^{k}\{X_{n_i}\in B_i\cap C\}\cap\bigcap_{j=k+1}^{l}\{X_{n_j}\in C\}\cap\bigcap_{n\in J(H)}\{X_n\notin C\}\right]$$

$$= \prod_{i=1}^{k}P[\{X\in B_i\cap C\}]\cdot\prod_{j=k+1}^{l}P[\{X\in C\}]\cdot\prod_{n\in J(H)}P[\{X\notin C\}]$$

$$= \prod_{i=1}^{k}P[\{X\in B_i\}|\{X\in C\}]\cdot\prod_{j=1}^{l}P[\{X\in C\}]\cdot\prod_{n\in J(H)}P[\{X\notin C\}]$$

$$= \prod_{i=1}^{k}P[\{X\in B_i\}|\{X\in C\}]\cdot\eta^{l}(1-\eta)^{n_i-l}\,,$$

as was to be shown.   □

6.2.5 Theorem (Thinned Claim Size Process). *The sequence $\{X'_n\}_{n\in\mathbf{N}}$ is i. i. d. and satisfies*

$$P[\{X'\in B\}] = P[\{X\in B\}|\{X\in C\}]$$

for all $B\in\mathcal{B}(\mathbf{R})$.

Proof. Consider $k\in\mathbf{N}$ and $B_1,\ldots,B_k\in\mathcal{B}(\mathbf{R})$. By Lemmas 6.2.4 and 6.2.2, we have, for every sequence $\{n_j\}_{j\in\{1,\ldots,k\}}\in\mathcal{H}(k)$,

$$P\left[\bigcap_{i=1}^{k}\{X'_i\in B_i\}\cap\bigcap_{j=1}^{k}\{\nu_j=n_j\}\right] = \prod_{i=1}^{k}P[\{X\in B_i\}|\{X\in C\}]\cdot\eta^k(1-\eta)^{n_k-k}$$

$$= \prod_{i=1}^{k}P[\{X\in B_i\}|\{X\in C\}]\cdot P\left[\bigcap_{j=1}^{k}\{\nu_j=n_j\}\right].$$

By Corollary 6.2.3, summation over $\mathcal{H}(k)$ yields

$$P\left[\bigcap_{i=1}^{k}\{X'_i\in B_i\}\right] = \prod_{i=1}^{k}P[\{X\in B_i\}|\{X\in C\}].$$

The assertion follows.   □

We can now prove the main result of this section:

6.2.6 Theorem (Thinned Risk Process). *The pair $(\{N'_t\}_{t\in\mathbf{R}_+},\{X'_n\}_{n\in\mathbf{N}})$ is a risk process.*

Proof. By Theorems 6.2.1 and 6.2.5, we know that $\{N'_t\}_{t\in\mathbf{R}_+}$ is a claim number process and that the sequence $\{X'_n\}_{n\in\mathbf{N}}$ is i. i. d.

To prove that the pair $(\{N'_t\}_{t \in \mathbf{R}_+}, \{X'_n\}_{n \in \mathbf{N}})$ is independent, consider $m, n \in \mathbf{N}$, $B_1, \ldots, B_n \in \mathcal{B}(\mathbf{R})$, $t_0, t_1, \ldots, t_m \in \mathbf{R}_+$ such that $0 = t_0 < t_1 < \ldots < t_m$, and $k_0, k_1, \ldots, k_m \in \mathbf{N}_0$ such that $0 = k_0 \le k_1 \le \ldots \le k_m$.
Consider also $l_0, l_1, \ldots, l_m \in \mathbf{N}_0$ satisfying $0 = l_0 \le l_1 \le \ldots \le l_m$ as well as $k_j - k_{j-1} \le l_j - l_{j-1}$ for all $j \in \{1, \ldots, m\}$. Define $n_0 := 0$ and $l := \max\{n, k_m + 1\}$, and let $\mathcal{H}$ denote the collection of all sequences $\{n_j\}_{j \in \{1, \ldots, l\}} \in \mathcal{H}(l)$ satisfying $n_{k_j} \le l_j < n_{k_j+1}$ for all $j \in \{1, \ldots, m\}$. By Lemma 6.2.4 and Theorem 6.2.5, we have

$$P\left[\bigcap_{i=1}^{n}\{X'_i \in B_i\} \cap \bigcap_{j=1}^{m}\left\{\sum_{h=1}^{l_j}\chi_{\{X_h \in C\}} = k_j\right\}\right]$$

$$= P\left[\bigcap_{i=1}^{n}\{X'_i \in B_i\} \cap \bigcap_{j=1}^{m}\{\nu_{k_j} \le l_j < \nu_{k_j+1}\}\right]$$

$$= \sum_{H \in \mathcal{H}} P\left[\bigcap_{i=1}^{n}\{X'_i \in B_i\} \cap \bigcap_{j=1}^{l}\{\nu_j = n_j\}\right]$$

$$= \sum_{H \in \mathcal{H}}\left(\prod_{i=1}^{n} P[\{X \in B_i\}|\{X \in C\}]\right)\eta^l(1-\eta)^{n_l-l}$$

$$= \sum_{H \in \mathcal{H}} P\left[\bigcap_{i=1}^{n}\{X'_i \in B_i\}\right]\eta^l(1-\eta)^{n_l-l}$$

$$= P\left[\bigcap_{i=1}^{n}\{X'_i \in B_i\}\right] \cdot \sum_{H \in \mathcal{H}}\eta^l(1-\eta)^{n_l-l},$$

hence

$$P\left[\bigcap_{i=1}^{n}\{X'_i \in B_i\} \cap \bigcap_{j=1}^{m}\{N'_{t_j} = k_j\} \cap \bigcap_{j=1}^{m}\{N_{t_j} = l_j\}\right]$$

$$= P\left[\bigcap_{i=1}^{n}\{X'_i \in B_i\} \cap \bigcap_{j=1}^{m}\left\{\sum_{h=1}^{N_{t_j}}\chi_{\{X_h \in C\}} = k_j\right\} \cap \bigcap_{j=1}^{m}\{N_{t_j} = l_j\}\right]$$

$$= P\left[\bigcap_{i=1}^{n}\{X'_i \in B_i\} \cap \bigcap_{j=1}^{m}\left\{\sum_{h=1}^{l_j}\chi_{\{X_h \in C\}} = k_j\right\} \cap \bigcap_{j=1}^{m}\{N_{t_j} = l_j\}\right]$$

$$= P\left[\bigcap_{i=1}^{n}\{X'_i \in B_i\} \cap \bigcap_{j=1}^{m}\left\{\sum_{h=1}^{l_j}\chi_{\{X_h \in C\}} = k_j\right\}\right] \cdot P\left[\bigcap_{j=1}^{m}\{N_{t_j} = l_j\}\right]$$

$$= P\left[\bigcap_{i=1}^{n}\{X'_i \in B_i\}\right] \cdot \sum_{H \in \mathcal{H}}\eta^l(1-\eta)^{n_l-l} \cdot P\left[\bigcap_{j=1}^{m}\{N_{t_j} = l_j\}\right],$$

and thus

$$P\left[\bigcap_{i=1}^{n}\{X_i' \in B_i\} \cap \bigcap_{j=1}^{m}\{N_{t_j}' = k_j\} \cap \bigcap_{j=1}^{m}\{N_{t_j} = l_j\}\right]$$

$$= P\left[\bigcap_{i=1}^{n}\{X_i' \in B_i\}\right] \cdot P\left[\bigcap_{j=1}^{m}\{N_{t_j}' = k_j\} \cap \bigcap_{j=1}^{m}\{N_{t_j} = l_j\}\right] .$$

Summation yields

$$P\left[\bigcap_{i=1}^{n}\{X_i' \in B_i\} \cap \bigcap_{j=1}^{m}\{N_{t_j}' = k_j\}\right] = P\left[\bigcap_{i=1}^{n}\{X_i' \in B_i\}\right] \cdot P\left[\bigcap_{j=1}^{m}\{N_{t_j}' = k_j\}\right] .$$

Therefore, the pair $(\{N_t'\}_{t \in \mathbf{R}_+}, \{X_n'\}_{n \in \mathbf{N}})$ is independent.   □

6.2.7 Corollary (Thinned Risk Process).   *Let* $h : \mathbf{R} \to \mathbf{R}$ *be a measurable function. Then the pair* $(\{N_t'\}_{t \in \mathbf{R}_+}, \{h(X_n')\}_{n \in \mathbf{N}})$ *is a risk process.*

This is immediate from Theorem 6.2.6.

As an application of Corollary 6.2.7, consider $c \in (0, \infty)$, let $C := (c, \infty)$, and define $h : \mathbf{R} \to \mathbf{R}$ by letting

$$h(x) := (x - c)^+ .$$

In excess of loss reinsurance, c is the priority of the direct insurer, and the reinsurer is concerned with the risk process $(\{N_t'\}_{t \in \mathbf{R}_+}, \{(X_n' - c)^+\}_{n \in \mathbf{N}})$.

More generally, consider $c, d \in (0, \infty)$, let $C := (c, \infty)$, and define $h : \mathbf{R} \to \mathbf{R}$ by letting

$$h(x) := (x - c)^+ \wedge d .$$

In this case, the reinsurer is not willing to pay more than d for a claim exceeding c and covers only the *layer* $(c, c+d]$; he is thus concerned with the risk process $(\{N_t'\}_{t \in \mathbf{R}_+}, \{(X_n' - c)^+ \wedge d\}_{n \in \mathbf{N}})$.

Problems

6.2.A   The sequence $\{\nu_l\}_{l \in \mathbf{N}_0}$ is a claim arrival process (in discrete time) satisfying

$$P_{\nu_l} = \mathbf{Geo}(l, \eta)$$

for all $l \in \mathbf{N}$. Moreover, the claim interarrival times are i.i.d. Study also the claim number process induced by the claim arrival process $\{\nu_l\}_{l \in \mathbf{N}_0}$.

6.2.B   For $c \in (0, \infty)$ and $C := (c, \infty)$, compute the distribution of X' for some specific choices of the distribution of X.

6.2.C   **Discrete Time Model:** The risk process $(\{N_l\}_{l \in \mathbf{N}_0}, \{X_n\}_{n \in \mathbf{N}})$ is a *binomial risk process* if the claim number process $\{N_l\}_{l \in \mathbf{N}_0}$ is a binomial process. Study the problem of thinning for a binomial risk process.

6.3   Decomposition of a Poisson Risk Process

Throughout this section, let $(\{N_t\}_{t\in\mathbf{R}_+}, \{X_n\}_{n\in\mathbf{N}})$ be a risk process and consider $C \in \mathcal{B}(\mathbf{R})$. Define

$$\eta := P[\{X\in C\}].$$

We assume that the probability of explosion is equal to zero and that $\eta \in (0,1)$.

Let us first consider the thinned claim number processes generated by the claims with claim size in C or in $\overline{C}$, respectively.

For all $t \in \mathbf{R}_+$, define

$$N_t' := \sum_{n=1}^{N_t} \chi_{\{X_n\in C\}}$$

and

$$N_t'' := \sum_{n=1}^{N_t} \chi_{\{X_n\in \overline{C}\}}.$$

By Theorem 6.2.1, $\{N_t'\}_{t\in\mathbf{R}_+}$ and $\{N_t''\}_{t\in\mathbf{R}_+}$ are claim number processes.

The following result improves Corollary 5.1.5:

6.3.1 Theorem (Decomposition of a Poisson Process).   *Assume that* $\{N_t\}_{t\in\mathbf{R}_+}$ *is a Poisson process with parameter* α*. Then the claim number processes* $\{N_t'\}_{t\in\mathbf{R}_+}$ *and* $\{N_t''\}_{t\in\mathbf{R}_+}$ *are independent Poisson processes with parameters* $\alpha\eta$ *and* $\alpha(1-\eta)$*, respectively.*

Proof.   Consider $m \in \mathbf{N}$, $t_0, t_1, \ldots, t_m \in \mathbf{R}_+$ such that $0 = t_0 < t_1 < \ldots < t_m$, and $k_1', \ldots, k_m' \in \mathbf{N}_0$ and $k_1'', \ldots, k_m'' \in \mathbf{N}_0$.
For all $j \in \{1, \ldots, m\}$, define $k_j := k_j' + k_j''$ and $n_j := \sum_{i=1}^{j} k_i$. Then we have

$$P\left[\bigcap_{j=1}^{m}\{N_{t_j}' - N_{t_{j-1}}' = k_j'\}\cap\{N_{t_j}'' - N_{t_{j-1}}'' = k_j''\}\,\middle|\,\bigcap_{j=1}^{m}\{N_{t_j} - N_{t_{j-1}} = k_j\}\right]$$

$$= P\left[\bigcap_{j=1}^{m}\left\{\sum_{h=N_{t_{j-1}}+1}^{N_{t_j}}\chi_{\{X_h\in C\}} = k_j'\right\}\cap\left\{\sum_{h=N_{t_{j-1}}+1}^{N_{t_j}}\chi_{\{X_h\in\overline{C}\}} = k_j''\right\}\,\middle|\,\bigcap_{j=1}^{m}\{N_{t_j} = n_j\}\right]$$

$$= P\left[\bigcap_{j=1}^{m}\left\{\sum_{h=n_{j-1}+1}^{n_j}\chi_{\{X_h\in C\}} = k_j'\right\}\cap\left\{\sum_{h=n_{j-1}+1}^{n_j}\chi_{\{X_h\in\overline{C}\}} = k_j''\right\}\,\middle|\,\bigcap_{j=1}^{m}\{N_{t_j} = n_j\}\right]$$

$$= P\left[\bigcap_{j=1}^{m}\left\{\sum_{h=n_{j-1}+1}^{n_j}\chi_{\{X_h\in C\}}=k_j'\right\}\cap\left\{\sum_{h=n_{j-1}+1}^{n_j}\chi_{\{X_h\in\overline{C}\}}=k_j''\right\}\right]$$

$$= \prod_{j=1}^{m}\binom{k_j}{k_j'}\eta^{k_j'}(1-\eta)^{k_j''}$$

as well as

$$P\left[\bigcap_{j=1}^{m}\{N_{t_j}-N_{t_{j-1}}=k_j\}\right] = \prod_{j=1}^{m}P[\{N_{t_j}-N_{t_{j-1}}=k_j\}]$$

$$= \prod_{j=1}^{m}e^{-\alpha(t_j-t_{j-1})}\frac{(\alpha(t_j-t_{j-1}))^{k_j}}{k_j!},$$

and hence

$$P\left[\bigcap_{j=1}^{m}\{N_{t_j}'-N_{t_{j-1}}'=k_j'\}\cap\{N_{t_j}''-N_{t_{j-1}}''=k_j''\}\right]$$

$$= P\left[\bigcap_{j=1}^{m}\{N_{t_j}'-N_{t_{j-1}}'=k_j'\}\cap\{N_{t_j}''-N_{t_{j-1}}''=k_j''\}\left|\bigcap_{j=1}^{m}\{N_{t_j}-N_{t_{j-1}}=k_j\}\right.\right]$$

$$\cdot P\left[\bigcap_{j=1}^{m}\{N_{t_j}-N_{t_{j-1}}=k_j\}\right]$$

$$= \prod_{j=1}^{m}\binom{k_j}{k_j'}\eta^{k_j'}(1-\eta)^{k_j''}\cdot\prod_{j=1}^{m}e^{-\alpha(t_j-t_{j-1})}\frac{(\alpha(t_j-t_{j-1}))^{k_j}}{k_j!}$$

$$= \prod_{j=1}^{m}e^{-\alpha\eta(t_j-t_{j-1})}\frac{(\alpha\eta(t_j-t_{j-1}))^{k_j'}}{k_j'!}\cdot\prod_{j=1}^{m}e^{-\alpha(1-\eta)(t_j-t_{j-1})}\frac{(\alpha(1-\eta)(t_j-t_{j-1}))^{k_j''}}{k_j''!}.$$

This implies that $\{N_t'\}_{t\in\mathbf{R}_+}$ and $\{N_t''\}_{t\in\mathbf{R}_+}$ are Poisson processes with parameters $\alpha\eta$ and $\alpha(1-\eta)$, respectively, and that $\{N_t'\}_{t\in\mathbf{R}_+}$ and $\{N_t''\}_{t\in\mathbf{R}_+}$ are independent. $\square$

Let us now consider the thinned claim size processes.

Let $\nu_0':=0$ and $\nu_0'':=0$. For all $l\in\mathbf{N}$, define

$$\nu_l' := \inf\{n\in\mathbf{N}\mid\nu_{l-1}'<n,\ X_n\in C\}$$

and

$$\nu_l'' := \inf\{n\in\mathbf{N}\mid\nu_{l-1}''<n,\ X_n\in\overline{C}\}.$$

For $m,n\in\mathbf{N}_0$ such that $m+n\in\mathbf{N}$, let $\mathcal{D}'(m,n)$ denote the collection of all pairs of strictly increasing sequences $\{m_i\}_{i\in\{1,\dots,k\}}\subseteq\mathbf{N}$ and $\{n_j\}_{j\in\{1,\dots,l\}}\subseteq\mathbf{N}$ satisfying

$k = m$ and $l \geq n$ as well as $n_l < m_k = k + l$ (such that one of these sequences may be empty and the union of these sequences is $\{1, \ldots, k + l\}$); similarly, let $\mathcal{D}''(m, n)$ denote the collection of all pairs of strictly increasing sequences $\{m_i\}_{i \in \{1, \ldots, k\}} \subseteq \mathbf{N}$ and $\{n_j\}_{j \in \{1, \ldots, l\}} \subseteq \mathbf{N}$ satisfying $l = n$ and $k \geq m$ as well as $m_k < n_l = k + l$, and define

$$\mathcal{D}(m, n) := \mathcal{D}'(m, n) + \mathcal{D}''(m, n).$$

The collections $\mathcal{D}(m, n)$ correspond to the collections $\mathcal{H}(l)$ considered in the previous section on thinning.

6.3.2 Lemma. *The identities*

$$\bigcap_{i=1}^{k} \{\nu_i' = m_i\} \cap \bigcap_{j=1}^{l} \{\nu_j'' = n_j\} = \bigcap_{i=1}^{k} \{X_{m_i} \in C\} \cap \bigcap_{j=1}^{l} \{X_{n_j} \in \overline{C}\}$$

and

$$P\left[\bigcap_{i=1}^{k} \{\nu_i' = m_i\} \cap \bigcap_{j=1}^{l} \{\nu_j'' = n_j\} \right] = \eta^k (1 - \eta)^l$$

hold for all $m, n \in \mathbf{N}_0$ *such that* $m + n \in \mathbf{N}$ *and for all* $(\{m_i\}_{i \in \{1, \ldots, k\}}, \{n_j\}_{j \in \{1, \ldots, l\}}) \in \mathcal{D}(m, n)$.

It is clear that, for every choice of $m, n \in \mathbf{N}_0$ such that $m + n \in \mathbf{N}$, the family $\{\bigcap_{i=1}^{k} \{\nu_i' = m_i\} \cap \bigcap_{j=1}^{l} \{\nu_j'' = n_j\}\}_{D \in \mathcal{D}(m,n)}$ is disjoint; the following lemma shows that it is, up to a null set, even a partition of Ω:

6.3.3 Corollary. *The identity*

$$\sum_{D \in \mathcal{D}(m,n)} P\left[\bigcap_{i=1}^{k} \{\nu_i' = m_i\} \cap \bigcap_{j=1}^{l} \{\nu_j'' = n_j\} \right] = 1$$

holds for all $m, n \in \mathbf{N}_0$ *such that* $m + n \in \mathbf{N}$.

Proof. For $m, n \in \mathbf{N}_0$ such that $m + n = 1$, the identity

$$\sum_{D \in \mathcal{D}(m,n)} P\left[\bigcap_{i=1}^{k} \{\nu_i' = m_i\} \cap \bigcap_{j=1}^{l} \{\nu_j'' = n_j\} \right] = 1$$

follows from Corollary 6.2.3.

For $m, n \in \mathbf{N}$, we split $\mathcal{D}(m, n)$ into two parts: Let $\mathcal{D}_1(m, n)$ denote the collection of all pairs $(\{m_i\}_{i \in \{1, \ldots, k\}}, \{n_j\}_{j \in \{1, \ldots, l\}}) \in \mathcal{D}(m, n)$ satisfying $m_1 = 1$, and let $\mathcal{D}_2(m, n)$ denote the collection of all pairs $(\{m_i\}_{i \in \{1, \ldots, k\}}, \{n_j\}_{j \in \{1, \ldots, l\}}) \in \mathcal{D}(m, n)$ satisfying $n_1 = 1$. Then we have

$$\mathcal{D}(m, n) = \mathcal{D}_1(m, n) + \mathcal{D}_2(m, n).$$

Furthermore, there are obvious bijections between $\mathcal{D}_1(m, n)$ and $\mathcal{D}(m - 1, n)$ and between $\mathcal{D}_2(m, n)$ and $\mathcal{D}(m, n - 1)$.

Using Lemma 6.3.2, the assertion now follows by induction over $m + n \in \mathbf{N}$. □

By Corollary 6.2.3, each ν'_n and each ν''_n is finite. For all $n \in \mathbf{N}$, define

$$X'_n \; := \; \sum_{k=1}^{\infty} \chi_{\{\nu'_n = k\}} X_k$$

and

$$X''_n \; := \; \sum_{k=1}^{\infty} \chi_{\{\nu''_n = k\}} X_k \, .$$

Then we have $\sigma(\{X'_n\}_{n \in \mathbf{N}}) \cup \sigma(\{X''_n\}_{n \in \mathbf{N}}) \subseteq \sigma(\{X_n\}_{n \in \mathbf{N}})$.

6.3.4 Lemma. *The identity*

$$P\left[\bigcap_{h=1}^{n} \{X'_h \in B'_h\} \cap \bigcap_{h=1}^{n} \{X''_h \in B''_h\} \cap \bigcap_{i=1}^{k} \{\nu'_i = m_i\} \cap \bigcap_{j=1}^{l} \{\nu''_j = n_j\}\right]$$

$$= \; \prod_{h=1}^{n} P[\{X'_h \in B'_h\}] \cdot \prod_{h=1}^{n} P[\{X''_h \in B''_h\}] \cdot P\left[\bigcap_{i=1}^{k} \{\nu'_i = m_i\} \cap \bigcap_{j=1}^{l} \{\nu''_j = n_j\}\right]$$

holds for all $n \in \mathbf{N}$, for all $B'_1, \ldots, B'_n \in \mathcal{B}(\mathbf{R})$ and $B''_1, \ldots, B''_n \in \mathcal{B}(\mathbf{R})$, and for every pair $(\{m_i\}_{i \in \{1,\ldots,k\}}, \{n_j\}_{j \in \{1,\ldots,l\}}) \in \mathcal{D}(n, n)$.

Proof. By Lemma 6.3.2 and Theorem 6.2.5, we have

$$P\left[\bigcap_{h=1}^{n} \{X'_h \in B'_h\} \cap \bigcap_{h=1}^{n} \{X''_h \in B''_h\} \cap \bigcap_{i=1}^{k} \{\nu'_i = m_i\} \cap \bigcap_{j=1}^{l} \{\nu''_j = n_j\}\right]$$

$$= \; P\left[\bigcap_{h=1}^{n} \{X_{m_i} \in B'_h\} \cap \bigcap_{h=1}^{n} \{X_{n_j} \in B''_h\} \cap \bigcap_{i=1}^{k} \{\nu'_i = m_i\} \cap \bigcap_{j=1}^{l} \{\nu''_j = n_j\}\right]$$

$$= \; P\left[\bigcap_{h=1}^{n} \{X_{m_i} \in B'_h\} \cap \bigcap_{h=1}^{n} \{X_{n_j} \in B''_h\} \cap \bigcap_{i=1}^{k} \{X_{m_i} \in C\} \cap \bigcap_{j=1}^{l} \{X_{n_j} \in \overline{C}\}\right]$$

$$= \; P\left[\bigcap_{h=1}^{n} \{X_{m_i} \in B'_h \cap C\} \cap \bigcap_{h=1}^{n} \{X_{n_j} \in B''_h \cap \overline{C}\} \cap \bigcap_{i=n+1}^{k} \{X_{m_i} \in C\} \cap \bigcap_{j=n+1}^{l} \{X_{n_j} \in \overline{C}\}\right]$$

$$= \; \prod_{h=1}^{n} P[\{X \in B'_h \cap C\}] \cdot \prod_{h=1}^{n} P[\{X \in B''_h \cap \overline{C}\}]$$

$$\cdot \prod_{i=n+1}^{k} P[\{X \in C\}] \cdot \prod_{j=n+1}^{l} P[\{X \in \overline{C}\}]$$

$$= \; \prod_{h=1}^{n} P[\{X \in B'_h\} | \{X \in C\}] \cdot \prod_{h=1}^{n} P[\{X \in B''_h\} | \{X \in \overline{C}\}]$$

$$\cdot \prod_{i=1}^{k} P[\{X \in C\}] \cdot \prod_{j=1}^{l} P[\{X \in \overline{C}\}]$$

$$= \prod_{h=1}^{n} P[\{X_h' \in B_h'\}] \cdot \prod_{h=1}^{n} P[\{X_h'' \in B_h''\}] \cdot \eta^k (1-\eta)^l$$

$$= \prod_{h=1}^{n} P[\{X_h' \in B_h'\}] \cdot \prod_{h=1}^{n} P[\{X_h'' \in B_h''\}] \cdot P\left[\bigcap_{i=1}^{k} \{\nu_i' = m_i\} \cap \bigcap_{j=1}^{l} \{\nu_j'' = n_j\}\right],$$

as was to be shown. □

6.3.5 Theorem (Decomposition of a Claim Size Process). *The claim size processes $\{X_n'\}_{n \in \mathbf{N}}$ and $\{X_n''\}_{n \in \mathbf{N}}$ are independent.*

Proof. Consider $n \in \mathbf{N}$, $B_1', \ldots, B_n' \in \mathcal{B}(\mathbf{R})$, and $B_1'', \ldots, B_n'' \in \mathcal{B}(\mathbf{R})$. By Lemma 6.3.4, we have, for every pair $(\{m_i\}_{i \in \{1,\ldots,k\}}, \{n_j\}_{j \in \{1,\ldots,l\}}) \in \mathcal{D}(n,n)$,

$$P\left[\bigcap_{h=1}^{n} \{X_h' \in B_h'\} \cap \bigcap_{h=1}^{n} \{X_h'' \in B_h''\} \cap \bigcap_{i=1}^{k} \{\nu_i' = m_i\} \cap \bigcap_{j=1}^{l} \{\nu_j'' = n_j\}\right]$$

$$= \prod_{h=1}^{n} P[\{X_h' \in B_h'\}] \cdot \prod_{h=1}^{n} P[\{X_h'' \in B_h''\}] \cdot P\left[\bigcap_{i=1}^{k} \{\nu_i' = m_i\} \cap \bigcap_{j=1}^{l} \{\nu_j'' = n_j\}\right].$$

By Corollary 6.3.3, summation over $\mathcal{D}(n,n)$ yields

$$P\left[\bigcap_{h=1}^{n} \{X_h' \in B_h'\} \cap \bigcap_{h=1}^{n} \{X_h'' \in B_h''\}\right] = \prod_{h=1}^{n} P[\{X_h' \in B_h'\}] \cdot \prod_{h=1}^{n} P[\{X_h'' \in B_h''\}].$$

The assertion follows. □

The risk process $(\{N_t\}_{t \in \mathbf{R}_+}, \{X_n\}_{n \in \mathbf{N}})$ is a *Poisson risk process* if the claim number process $\{N_t\}_{t \in \mathbf{R}_+}$ is a Poisson process.

We can now prove the main result of this section:

6.3.6 Theorem (Decomposition of a Poisson Risk Process). *Assume that $(\{N_t\}_{t \in \mathbf{R}_+}, \{X_n\}_{n \in \mathbf{N}})$ is a Poisson risk process. Then $(\{N_t'\}_{t \in \mathbf{R}_+}, \{X_n'\}_{n \in \mathbf{N}})$ and $(\{N_t''\}_{t \in \mathbf{R}_+}, \{X_n''\}_{n \in \mathbf{N}})$ are independent Poisson risk processes.*

Proof. By Theorems 6.3.1 and 6.3.5, we know that $\{N_t'\}_{t \in \mathbf{R}_+}$ and $\{N_t''\}_{t \in \mathbf{R}_+}$ are independent Poisson processes and that $\{X_n'\}_{n \in \mathbf{N}}$ and $\{X_n''\}_{n \in \mathbf{N}}$ are independent claim size processes.

To prove that the σ–algebras $\sigma(\{N_t'\}_{t \in \mathbf{R}_+} \cup \{N_t''\}_{t \in \mathbf{R}_+})$ and $\sigma(\{X_n'\}_{n \in \mathbf{N}} \cup \{X_n''\}_{n \in \mathbf{N}})$ are independent, consider $m, n \in \mathbf{N}$, $B_1', \ldots, B_n' \in \mathcal{B}(\mathbf{R})$ and $B_1'', \ldots, B_n'' \in \mathcal{B}(\mathbf{R})$, as well as $t_0, t_1, \ldots, t_m \in \mathbf{R}_+$ such that $0 = t_0 < t_1 < \ldots < t_m$ and $k_1', \ldots, k_m' \in \mathbf{N}_0$ and $k_1'', \ldots, k_m'' \in \mathbf{N}_0$ such that $k_1' \leq \ldots \leq k_m'$ and $k_1'' \leq \ldots \leq k_m''$. For all $j \in \{1, \ldots, m\}$, define $k_j = k_j' + k_j''$.

Furthermore, let $p := \max\{n, k_m', k_m''\}$, and let $\mathcal{D}$ denote the collection of all pairs $(\{m_i\}_{i \in \{1,\ldots,k\}}, \{n_j\}_{j \in \{1,\ldots,l\}}) \in \mathcal{D}(p,p)$ satisfying $\max\{m_{k_j'}, n_{k_j''}\} = k_j$ for all $j \in \{1, \ldots, m\}$. By Lemma 6.3.4 and Theorem 6.3.5, we have

$$P\left[\bigcap_{h=1}^{n}\{X'_h \in B'_h\}\cap\{X''_h \in B''_h\} \cap \bigcap_{j=1}^{m}\left\{\sum_{r=1}^{k_j}\chi_{\{X_r\in C\}} = k'_j\right\}\right]$$

$$= \sum_{D\in\mathcal{D}} P\left[\bigcap_{h=1}^{n}\{X'_h \in B'_h\}\cap\{X''_h \in B''_h\} \cap \bigcap_{j=1}^{m}\{\max\{\nu'_{k'_j}, \nu''_{k''_j}\} = k_j\}\right]$$

$$= \sum_{D\in\mathcal{D}} P\left[\bigcap_{h=1}^{n}\{X'_h \in B'_h\}\cap\{X''_h \in B''_h\} \cap \bigcap_{i=1}^{k}\{\nu'_i = m_i\} \cap \bigcap_{j=1}^{l}\{\nu''_j = n_j\}\right].$$

$$= \sum_{D\in\mathcal{D}}\left(\prod_{h=1}^{n}P[\{X'_h \in B'_h\}]\cdot\prod_{h=1}^{n}P[\{X''_h \in B''_h\}]\cdot\eta^k(1-\eta)^l\right)$$

$$= \sum_{D\in\mathcal{D}} P\left[\bigcap_{h=1}^{n}\{X'_h \in B'_h\}\cap\{X''_h \in B''_h\}\right]\eta^k(1-\eta)^l$$

$$= P\left[\bigcap_{h=1}^{n}\{X'_h \in B'_h\}\cap\{X''_h \in B''_h\}\right]\cdot\sum_{D\in\mathcal{D}}\eta^k(1-\eta)^l\,,$$

hence

$$P\left[\bigcap_{h=1}^{n}\{X'_h \in B'_h\}\cap\{X''_h \in B''_h\} \cap \bigcap_{j=1}^{m}\{N'_{t_j} = k'_j\}\cap\{N''_{t_j} = k''_j\}\right]$$

$$= P\left[\bigcap_{h=1}^{n}\{X'_h \in B'_h\}\cap\{X''_h \in B''_h\} \cap \bigcap_{j=1}^{m}\{N'_{t_j} = k'_j\}\cap\{N_{t_j} = k_j\}\right]$$

$$= P\left[\bigcap_{h=1}^{n}\{X'_h \in B'_h\}\cap\{X''_h \in B''_h\} \cap \bigcap_{j=1}^{m}\left\{\sum_{r=1}^{N_{t_j}}\chi_{\{X_r\in C\}} = k'_j\right\}\cap\{N_{t_j} = k_j\}\right]$$

$$= P\left[\bigcap_{h=1}^{n}\{X'_h \in B'_h\}\cap\{X''_h \in B''_h\} \cap \bigcap_{j=1}^{m}\left\{\sum_{r=1}^{k_j}\chi_{\{X_r\in C\}} = k'_j\right\}\cap\{N_{t_j} = k_j\}\right]$$

$$= P\left[\bigcap_{h=1}^{n}\{X'_h \in B'_h\}\cap\{X''_h \in B''_h\} \cap \bigcap_{j=1}^{m}\left\{\sum_{r=1}^{k_j}\chi_{\{X_r\in C\}} = k'_j\right\}\cdot P\left[\bigcap_{j=1}^{m}\{N_{t_j} = k_j\}\right]\right]$$

$$= P\left[\bigcap_{h=1}^{n}\{X'_h \in B'_h\}\cap\{X''_h \in B''_h\}\right]\cdot\sum_{D\in\mathcal{D}}\eta^k(1-\eta)^l\cdot P\left[\bigcap_{j=1}^{m}\{N_{t_j} = k_j\}\right]\,,$$

and thus

$$P\left[\bigcap_{h=1}^{n}\{X'_h \in B'_h\}\cap\{X''_h \in B''_h\} \cap \bigcap_{j=1}^{m}\{N'_{t_j} = k'_j\}\cap\{N''_{t_j} = k''_j\}\right]$$

$$= P\left[\bigcap_{h=1}^{n}\{X'_h \in B'_h\}\cap\{X''_h \in B''_h\}\right]\cdot P\left[\bigcap_{j=1}^{m}\{N'_{t_j} = k'_j\}\cap\{N''_{t_j} = k''_j\}\right].$$

Therefore, the σ–algebras $\sigma(\{N_t'\}_{t\in\mathbf{R}_+} \cup \{N_t''\}_{t\in\mathbf{R}_+})$ and $\sigma(\{X_n'\}_{n\in\mathbf{N}} \cup \{X_n''\}_{n\in\mathbf{N}})$ are independent.
The assertion follows. $\square$

Let $\{S_t'\}_{t\in\mathbf{R}_+}$ and $\{S_t''\}_{t\in\mathbf{R}_+}$ denote the aggregate claims processes induced by the risk processes $(\{N_t'\}_{t\in\mathbf{R}_+}, \{X_n'\}_{n\in\mathbf{N}})$ and $(\{N_t''\}_{t\in\mathbf{R}_+}, \{X_n''\}_{n\in\mathbf{N}})$, respectively. In the case where $(\{N_t\}_{t\in\mathbf{R}_+}, \{X_n\}_{n\in\mathbf{N}})$ is a Poisson risk process, one should expect that the sum of the aggregate claims processes $\{S_t'\}_{t\in\mathbf{R}_+}$ and $\{S_t''\}_{t\in\mathbf{R}_+}$ agrees with the aggregate claims process $\{S_t\}_{t\in\mathbf{R}_+}$. The following result asserts that this is indeed true:

6.3.7 Theorem.   *Assume that* $(\{N_t\}_{t\in\mathbf{R}_+}, \{X_n\}_{n\in\mathbf{N}})$ *is a Poisson risk process. Then the aggregate claims processes* $\{S_t'\}_{t\in\mathbf{R}_+}$ *and* $\{S_t''\}_{t\in\mathbf{R}_+}$ *are independent, and the identity*

$$S_t \;=\; S_t' + S_t''$$

holds for all $t \in \mathbf{R}_+$.

Proof.   By Theorem 6.3.6, the aggregate claims processes $\{S_t'\}_{t\in\mathbf{R}_+}$ and $\{S_t''\}_{t\in\mathbf{R}_+}$ are independent.
For all $\omega \in \{N_t' = 0\} \cap \{N_t'' = 0\}$, we clearly have

$$S_t'(\omega) + S_t''(\omega) \;=\; S_t(\omega) \,.$$

Consider now $k', k'' \in \mathbf{N}_0$ such that $k' + k'' \in \mathbf{N}$. Note that $\max\{\nu_{k'}', \nu_{k''}''\} \geq k'+k''$, and that

$$\{N_t' = k'\} \cap \{N_t'' = k''\} \;=\; \{\max\{\nu_{k'}', \nu_{k''}''\} = k'+k''\} \cap \{N_t = k'+k''\} \,.$$

Thus, for $(\{m_i\}_{i\in\{1,\dots,k\}}, \{n_j\}_{j\in\{1,\dots,l\}}) \in \mathcal{D}(k', k'')$ satisfying $\max\{m_{k'}, n_{k''}\} > k'+k''$ the set $\{N_t' = k'\} \cap \{N_t'' = k''\} \cap \bigcap_{i=1}^{k}\{\nu_i' = m_i\} \cap \bigcap_{j=1}^{l}\{\nu_j'' = n_j\}$ is empty, and for $(\{m_i\}_{i\in\{1,\dots,k\}}, \{n_j\}_{j\in\{1,\dots,l\}}) \in \mathcal{D}(k', k'')$ satisfying $\max\{m_{k'}, n_{k''}\} = k'+k''$ we have, for all $\omega \in \{N_t' = k'\} \cap \{N_t'' = k''\} \cap \bigcap_{i=1}^{k}\{\nu_i' = m_i\} \cap \bigcap_{j=1}^{l}\{\nu_j'' = n_j\}$,

$$
\begin{aligned}
S_t'(\omega) + S_t''(\omega) &= \sum_{i=1}^{N_t'(\omega)} X_i'(\omega) + \sum_{j=1}^{N_t''(\omega)} X_j''(\omega) \\
&= \sum_{i=1}^{k'} X_i'(\omega) + \sum_{j=1}^{k''} X_j''(\omega) \\
&= \sum_{i=1}^{k'} X_{m_i}(\omega) + \sum_{j=1}^{k''} X_{n_j}(\omega) \\
&= \sum_{h=1}^{k'+k''} X_h(\omega)
\end{aligned}
$$

$$= \sum_{h=1}^{N_t} X_h(\omega)$$

$$= S_t(\omega) \,.$$

This yields

$$S_t'(\omega) + S_t''(\omega) \;=\; S_t(\omega)$$

for all $\omega \in \{N_t' = k'\} \cap \{N_t'' = k''\}$.
We conclude that

$$S_t' + S_t'' \;=\; S_t \,,$$

as was to be shown.   □

Problems

6.3.A   Let $(\{N_t\}_{t \in \mathbf{R}_+}, \{X_n\}_{n \in \mathbf{N}})$ be a Poisson risk process satisfying $P_X = \mathbf{Exp}(\beta)$,
and let $C := (c, \infty)$ for some $c \in (0, \infty)$.
Compute the distributions of X' and X'', compare the expectation, variance and
coefficient of variation of X' and of X'' with the corresponding quantities of X,
and compute these quantities for S_t', S_t'', and S_t (see Problem 5.2.G).

6.3.B   Extend the results of this section to the decomposition of a Poisson risk process
into more than two Poisson risk processes.

6.3.C   Let $(\{N_t\}_{t \in \mathbf{R}_+}, \{X_n\}_{n \in \mathbf{N}})$ be a Poisson risk process satisfying $P_X[\{1, \ldots, m\}] = 1$
for some $m \in \mathbf{N}$. For all $j \in \{1, \ldots, m\}$ and $t \in \mathbf{R}_+$, define

$$N_t^{(j)} := \sum_{n=1}^{N_t} \chi_{\{X_n = j\}} \,.$$

Then the claim number processes $\{N_t^{(1)}\}_{t \in \mathbf{R}_+}, \ldots, \{N_t^{(m)}\}_{t \in \mathbf{R}_+}$ are independent,
and the identity

$$S_t \;=\; \sum_{j=1}^{m} j \, N_t^{(j)}$$

holds for all $t \in \mathbf{R}_+$.

6.3.D   Extend the results of this section to the case where the claim number process is
an inhomogeneous Poisson process.

6.3.E   **Discrete Time Model:** Study the decomposition of a binomial risk process.

6.3.F   **Discrete Time Model:** Let $(\{N_l\}_{l \in \mathbf{N}_0}, \{X_n\}_{n \in \mathbf{N}})$ be a binomial risk process
satisfying $P_X = \mathbf{Geo}(\eta)$, and let $C := (m, \infty)$ for some $m \in \mathbf{N}$.
Compute the distributions of X' and X'', compare the expectation, variance and
coefficient of variation of X' and of X'' with the corresponding quantities of X,
and compute these quantities for S_l', S_l'', and S_l.

6.4 Superposition of Poisson Risk Processes

Throughout this section, let $(\{N_t'\}_{t\in\mathbf{R}_+}, \{X_n'\}_{n\in\mathbf{N}})$ and $(\{N_t''\}_{t\in\mathbf{R}_+}, \{X_n''\}_{n\in\mathbf{N}})$ be risk processes, let $\{T_n'\}_{n\in\mathbf{N}_0}$ and $\{T_n''\}_{n\in\mathbf{N}_0}$ denote the claim arrival processes induced by the claim number processes $\{N_t'\}_{t\in\mathbf{R}_+}$ and $\{N_t''\}_{t\in\mathbf{R}_+}$, and let $\{W_n'\}_{n\in\mathbf{N}}$ and $\{W_n''\}_{n\in\mathbf{N}}$ denote the claim interarrival processes induced by the claim arrival processes $\{T_n'\}_{n\in\mathbf{N}_0}$ and $\{T_n''\}_{n\in\mathbf{N}_0}$, respectively.

We assume that the risk processes $(\{N_t'\}_{t\in\mathbf{R}_+}, \{X_n'\}_{n\in\mathbf{N}})$ and $(\{N_t''\}_{t\in\mathbf{R}_+}, \{X_n''\}_{n\in\mathbf{N}})$ are independent, that their exceptional null sets are empty, and that the claim number processes $\{N_t'\}_{t\in\mathbf{R}_+}$ and $\{N_t''\}_{t\in\mathbf{R}_+}$ are Poisson processes with parameters α' and α'', respectively.

For all $t \in \mathbf{R}_+$, define

$$N_t := N_t' + N_t'' \,.$$

The process $\{N_t\}_{t\in\mathbf{R}_+}$ is said to be the *superposition* of the Poisson processes $\{N_t'\}_{t\in\mathbf{R}_+}$ and $\{N_t''\}_{t\in\mathbf{R}_+}$.

The following result shows that the class of all Poisson processes is stable under superposition:

6.4.1 Theorem (Superposition of Poisson Processes). *The process $\{N_t\}_{t\in\mathbf{R}_+}$ is a Poisson process with parameter $\alpha' + \alpha''$.*

Proof. We first show that $\{N_t\}_{t\in\mathbf{R}_+}$ is a claim number process, and we then prove that it is indeed a Poisson process. To simplify the notation in this proof, let

$$\alpha := \alpha' + \alpha'' \,.$$

(1) Define

$$\Omega_N := \bigcup_{n',n''\in\mathbf{N}_0} \{T_{n'}' = T_{n''}''\} \,.$$

Since the claim number processes $\{N_t'\}_{t\in\mathbf{R}_+}$ and $\{N_t''\}_{t\in\mathbf{R}_+}$ are independent and since the distributions of their claim arrival times are absolutely continuous with respect to Lebesgue measure, we have

$$P[\Omega_N] = 0 \,.$$

It is now easy to see that $\{N_t\}_{t\in\mathbf{R}_+}$ is a claim number process with exceptional null set Ω_N.

(2) For all $t \in \mathbf{R}_+$, we have

$$\begin{aligned}
E[N_t] &= E[N_t'] + E[N_t''] \\
&= \alpha' t + \alpha'' t \\
&= \alpha t \,.
\end{aligned}$$

(3) Let $\{\mathcal{F}'_t\}_{t\in\mathbf{R}_+}$ and $\{\mathcal{F}''_t\}_{t\in\mathbf{R}_+}$ denote the canonical filtrations of $\{N'_t\}_{t\in\mathbf{R}_+}$ and $\{N''_t\}_{t\in\mathbf{R}_+}$, respectively, and let $\{\mathcal{F}_t\}_{t\in\mathbf{R}_+}$ denote the canonical filtration of $\{N_t\}_{t\in\mathbf{R}_+}$. Using independence of the pair $(\{N'_t\}_{t\in\mathbf{R}_+}, \{N''_t\}_{t\in\mathbf{R}_+})$ and the martingale property of the centered claim number processes $\{N'_t - \alpha't\}_{t\in\mathbf{R}_+}$ and $\{N''_t - \alpha''t\}_{t\in\mathbf{R}_+}$, which is given by Theorem 2.3.4, we see that the identity

$$
\begin{aligned}
&\int_{A'\cap A''} (N_{t+h} - N_t - \alpha h)\,dP \\
&= \int_{A'\cap A''} \Big((N'_{t+h}+N''_{t+h}) - (N'_t + N''_t) - (\alpha'+\alpha'')h \Big)\,dP \\
&= \int_{A'\cap A''} (N'_{t+h} - N'_t - \alpha'h)\,dP + \int_{A'\cap A''} (N''_{t+h} - N''_t - \alpha''h)\,dP \\
&= P[A''] \int_{A'} (N'_{t+h} - N'_t - \alpha'h)\,dP + P[A'] \int_{A''} (N''_{t+h} - N''_t - \alpha''h)\,dP \\
&= 0
\end{aligned}
$$

holds for all $t, h \in \mathbf{R}_+$ and for all $A' \in \mathcal{F}'_t$ and $A'' \in \mathcal{F}''_t$. Thus, letting

$$
\mathcal{E}_t := \{A'\cap A'' \mid A' \in \mathcal{F}'_t,\ A'' \in \mathcal{F}''_t\} \,,
$$

the previous identity yields

$$
\int_A (N_{t+h} - \alpha(t+h))\,dP = \int_A (N_t - \alpha t)\,dP
$$

for all $t, h \in \mathbf{R}_+$ and $A \in \mathcal{E}_t$. Since $\mathcal{E}_t$ is stable under intersection and satisfies $\mathcal{F}_t \subseteq \sigma(\mathcal{E}_t)$, we conclude that the identity

$$
\int_A (N_{t+h} - \alpha(t+h))\,dP = \int_A (N_t - \alpha t)\,dP
$$

holds for all $t, h \in \mathbf{R}_+$ and $A \in \mathcal{F}_t$. Therefore, the centered claim number process $\{N_t - \alpha t\}_{t\in\mathbf{R}_+}$ is a martingale, and it now follows from Theorem 2.3.4 that the claim number process $\{N_t\}_{t\in\mathbf{R}_+}$ is a Poisson process with parameter α. $\square$

Let $\{T_n\}_{n\in\mathbf{N}_0}$ denote the claim arrival process induced by the claim number process $\{N_t\}_{t\in\mathbf{R}_+}$ and let $\{W_n\}_{n\in\mathbf{N}}$ denote the claim interarrival process induced by $\{T_n\}_{n\in\mathbf{N}_0}$.

To avoid the annoying discussion of null sets, we assume henceforth that the exceptional null set of the claim number process $\{N_t\}_{t\in\mathbf{R}_+}$ is empty.

The following result shows that each of the distributions $P_{W'}$ and $P_{W''}$ has a density with respect to P_W:

6.4.2 Lemma. *The distributions $P_{W'}$ and $P_{W''}$ satisfy*

$$P_{W'} = \int \frac{\alpha'}{\alpha' + \alpha''} e^{\alpha'' w} \, dP_W(w)$$

and

$$P_{W''} = \int \frac{\alpha''}{\alpha' + \alpha''} e^{\alpha' w} \, dP_W(w) \, .$$

Proof. Since $P_{W'} = \mathbf{Exp}(\alpha')$ and $P_W = \mathbf{Exp}(\alpha' + \alpha'')$, we have

$$
\begin{aligned}
P_{W'} &= \int \alpha' e^{-\alpha' w} \chi_{(0,\infty)}(w) \, d\lambda(w) \\
&= \int \frac{\alpha'}{\alpha' + \alpha''} e^{\alpha'' w} (\alpha' + \alpha'') e^{-(\alpha' + \alpha'')w} \chi_{(0,\infty)}(w) \, d\lambda(w) \\
&= \int \frac{\alpha'}{\alpha' + \alpha''} e^{\alpha'' w} \, dP_W(w) \, ,
\end{aligned}
$$

which is the first identity. The second identity follows by symmetry. □

For $l \in \mathbf{N}$ and $k \in \{0, 1, \ldots, l\}$, let $\mathcal{C}(l, k)$ denote the collection of all pairs of strictly increasing sequences $\{m_i\}_{i \in \{1,\ldots,k\}} \subseteq \mathbf{N}$ and $\{n_j\}_{j \in \{1,\ldots,l-k\}} \subseteq \mathbf{N}$ with union $\{1, \ldots, l\}$ (such that one of these sequences may be empty).

For $l \in \mathbf{N}$, define

$$\mathcal{C}(l) = \sum_{k=0}^{l} \mathcal{C}(l, k) \, .$$

The collections $\mathcal{C}(l)$ correspond to the collections $\mathcal{H}(l)$ and $\mathcal{D}(m, n)$ considered in the preceding sections on thinning and decomposition.

For $l \in \mathbf{N}$, $k \in \{0, 1, \ldots, l\}$, and $C = (\{m_i\}_{i \in \{1,\ldots,k\}}, \{n_j\}_{j \in \{1,\ldots,l-k\}}) \in \mathcal{C}(l, k)$, let

$$A_C := \bigcap_{i \in \{1,\ldots,k\}} \{T_{m_i} = T_i'\} \cap \bigcap_{j \in \{1,\ldots,l-k\}} \{T_{n_j} = T_j''\} \, .$$

We have the following lemma:

6.4.3 Lemma. *The identity*

$$P[A_C] = \left(\frac{\alpha'}{\alpha' + \alpha''} \right)^k \left(\frac{\alpha''}{\alpha' + \alpha''} \right)^{l-k}$$

holds for all $l \in \mathbf{N}$ and $k \in \{0, 1, \ldots, l\}$ and for all $C \in \mathcal{C}(l, k)$.

Proof.   Consider $C = (\{m_i\}_{i \in \{1,\dots,k\}}, \{n_j\}_{j \in \{1,\dots,l-k\}}) \in \mathcal{C}(l,k)$. We prove several auxiliary results from which the assertion will follow by induction over $l \in \mathbf{N}$.

(1) *If $l = 1$ and $m_k = l$, then*

$$P[\{T_1 = T_1'\}] \;=\; \frac{\alpha'}{\alpha' + \alpha''} \,.$$

We have

$$\begin{aligned}
\{T_1 < T_1'\} \;&=\; \bigcup_{t \in \mathbf{Q}} \{T_1 \leq t < T_1'\} \\
&=\; \bigcup_{t \in \mathbf{Q}} \{N_t \geq 1\} \cap \{N_t' = 0\} \\
&=\; \bigcup_{t \in \mathbf{Q}} \{N_t'' \geq 1\} \cap \{N_t' = 0\} \\
&=\; \bigcup_{t \in \mathbf{Q}} \{T_1'' \leq t < T_1'\} \\
&=\; \{T_1'' < T_1'\}
\end{aligned}$$

as well as, by a similar argument,

$$\{T_1 > T_1'\} \;=\; \emptyset \,,$$

and thus

$$\begin{aligned}
\{T_1 = T_1'\} \;&=\; \{T_1' \leq T_1''\} \\
&=\; \{T_1' < T_1''\} \\
&=\; \{W_1' < W_1''\} \,.
\end{aligned}$$

(In the sequel, arguments of this type will be tacitly used at several occasions.) Now Lemma 6.4.2 yields

$$\begin{aligned}
P[\{T_1 = T_1'\}] \;&=\; P[\{W_1' < W_1''\}] \\
&=\; \int_{\mathbf{R}} P[\{w < W_1''\}] \, dP_{W_1'}(w) \\
&=\; \int_{\mathbf{R}} e^{-\alpha'' w} \, \frac{\alpha'}{\alpha' + \alpha''} \, e^{\alpha'' w} \, dP_W(w) \\
&=\; \frac{\alpha'}{\alpha' + \alpha''} \,,
\end{aligned}$$

as was to be shown.

(2) *If $l = 1$ and $n_{l-k} = l$, then*

$$P[\{T_1 = T_1''\}] \;=\; \frac{\alpha''}{\alpha' + \alpha''} \,.$$

This follows from (1) by symmetry.

(3) *If $l \geq 2$ and $m_k = l$, then*

$$P\left[\bigcap_{i \in \{1,\dots,k\}} \{T_{m_i} = T'_i\} \cap \bigcap_{j \in \{1,\dots,l-k\}} \{T_{n_j} = T''_j\}\right]$$

$$= \frac{\alpha'}{\alpha' + \alpha''} \cdot P\left[\bigcap_{i \in \{1,\dots,k-1\}} \{T_{m_i} = T'_i\} \cap \bigcap_{j \in \{1,\dots,l-k\}} \{T_{n_j} = T''_j\}\right].$$

This means that elimination of the event $\{T_{m_k} = T'_k\}$ produces the factor $\alpha'/(\alpha'+\alpha'')$. To prove our claim, we consider separately the cases $m_{k-1} < l-1$ and $m_{k-1} = l-1$. Let us first consider the case $m_{k-1} < l-1$. For all $s, t \in (0, \infty)$ such that $s \leq t$, Lemma 6.4.2 yields

$$\int_{(t-s,\infty)} \int_{(s+v-t,\infty)} dP_{W''}(w)\, dP_{W'}(v) = \int_{(t-s,\infty)} e^{-\alpha''(s+v-t)}\, dP_{W'}(v)$$

$$= \int_{(t-s,\infty)} e^{-\alpha''(s+v-t)} \frac{\alpha'}{\alpha' + \alpha''} e^{\alpha''v}\, dP_W(v)$$

$$= \frac{\alpha'}{\alpha' + \alpha''} e^{-\alpha''(s-t)} \int_{(t-s,\infty)} dP_W(v)$$

$$= \frac{\alpha'}{\alpha' + \alpha''} e^{-\alpha''(s-t)} e^{-(\alpha'+\alpha'')(t-s)}$$

$$= \frac{\alpha'}{\alpha' + \alpha''} e^{-\alpha'(t-s)}$$

$$= \frac{\alpha'}{\alpha' + \alpha''} \int_{(t-s,\infty)} dP_{W'}(v) .$$

For integration variables $v_1, \dots, v_{k-1} \in \mathbf{R}$ and $w_1, \dots, w_{l-k} \in \mathbf{R}$, define

$$s := \sum_{h=1}^{k-1} v_h$$

and

$$t := \sum_{h=1}^{l-k} w_h .$$

With suitable domains of integration where these are not specified, the identity established before yields

$$P\left[\bigcap_{i \in \{1,\dots,k\}} \{T_{m_i} = T'_i\} \cap \bigcap_{j \in \{1,\dots,l-k\}} \{T_{n_j} = T''_j\}\right]$$

$$= P[\{\dots < T'_{k-1} \dots < T''_{l-k} < T'_k < T''_{l-k+1}\}]$$

$$= P[\{\dots < T'_{k-1} \dots < T''_{l-k} < T'_{k-1} + W'_k < T''_{l-k} + W''_{l-k+1}\}]$$

$$
= \ldots \int \ldots \iint_{(t-s,\infty)} \int_{(s+v-t,\infty)} dP_{W''_{l-k+1}}(w)\, dP_{W'_k}(v)\, dP_{W''_{l-k}}(w_{l-k}) \ldots dP_{W'_{k-1}}(v_{k-1}) \ldots
$$

$$
= \ldots \int \ldots \int \left(\frac{\alpha'}{\alpha'+\alpha''} \int_{(t-s,\infty)} dP_{W'_k}(v) \right) dP_{W''_{l-k}}(w_{l-k}) \ldots dP_{W'_{k-1}}(v_{k-1}) \ldots
$$

$$
= \frac{\alpha'}{\alpha'+\alpha''} \cdot \left(\ldots \int \ldots \iint_{(t-s,\infty)} dP_{W'_k}(v)\, dP_{W''_{l-k}}(w_{l-k}) \ldots dP_{W'_{k-1}}(v_{k-1}) \ldots \right)
$$

$$
= \frac{\alpha'}{\alpha'+\alpha''} \cdot P[\{ \ldots < T'_{k-1} \ldots < T''_{l-k} < T'_{k-1} + W'_k \}]
$$

$$
= \frac{\alpha'}{\alpha'+\alpha''} \cdot P[\{ \ldots < T'_{k-1} \ldots < T''_{l-k} < T'_k \}]
$$

$$
= \frac{\alpha'}{\alpha'+\alpha''} \cdot P\left[\bigcap_{i \in \{1,\ldots,k-1\}} \{T_{m_i} = T'_i\} \cap \bigcap_{j \in \{1,\ldots,l-k\}} \{T_{n_j} = T''_j\} \right].
$$

Let us now consider the case $m_{k-1} = l-1$. For all $s,t \in (0,\infty)$ such that $t \leq s$, Lemma 6.4.2 yields

$$
\int_{(0,\infty)} \int_{(s+v-t,\infty)} dP_{W''}(w)\, dP_{W'}(v) = \int_{(0,\infty)} e^{-\alpha''(s+v-t)}\, dP_{W'}(v)
$$

$$
= \int_{(0,\infty)} e^{-\alpha''(s+v-t)} \frac{\alpha'}{\alpha'+\alpha''} e^{\alpha''v}\, dP_W(v)
$$

$$
= \frac{\alpha'}{\alpha'+\alpha''} e^{-\alpha''(s-t)}
$$

$$
= \frac{\alpha'}{\alpha'+\alpha''} \int_{(s-t,\infty)} dP_{W''}(w).
$$

This yields, as before,

$$
P\left[\bigcap_{i \in \{1,\ldots,k\}} \{T_{m_i} = T'_i\} \cap \bigcap_{j \in \{1,\ldots,l-k\}} \{T_{n_j} = T''_j\} \right]
$$

$$
= P[\{ \ldots < T''_{l-k} \ldots < T'_{k-1} < T'_k < T''_{l-k+1} \}]
$$

$$
= P[\{ \ldots < T''_{l-k} \ldots < T'_{k-1} < T'_{k-1} + W'_k < T''_{l-k} + W''_{l-k+1} \}]
$$

$$
= \ldots \int \ldots \iint_{(0,\infty)} \int_{(s+v-t,\infty)} dP_{W''_{l-k+1}}(w)\, dP_{W'_k}(v)\, dP_{W'_{k-1}}(v_{k-1}) \ldots dP_{W''_{l-k}}(w_{l-k}) \ldots
$$

$$
= \ldots \int \ldots \int \left(\frac{\alpha'}{\alpha'+\alpha''} \int_{(s-t,\infty)} dP_{W''_{l-k+1}}(v) \right) dP_{W'_{k-1}}(v_{k-1}) \ldots dP_{W''_{l-k}}(w_{l-k}) \ldots
$$

$$
= \frac{\alpha'}{\alpha'+\alpha''} \cdot \left(\ldots \int \ldots \iint_{(s-t,\infty)} dP_{W''_{l-k+1}}(v)\, dP_{W'_{k-1}}(v_{k-1}) \ldots dP_{W''_{l-k}}(w_{l-k}) \ldots \right)
$$

$$
= \frac{\alpha'}{\alpha' + \alpha''} \cdot P[\{\ldots < T''_{l-k} \cdots < T'_{k-1} < T''_{l-k} + W''_{l-k+1}\}]
$$

$$
= \frac{\alpha'}{\alpha' + \alpha''} \cdot P[\{\ldots < T''_{l-k} \cdots < T'_{k-1} < T''_{l-k+1}\}]
$$

$$
= \frac{\alpha'}{\alpha' + \alpha''} \cdot P\left[\bigcap_{i \in \{1,\ldots,k-1\}} \{T_{m_i} = T'_i\} \cap \bigcap_{j \in \{1,\ldots,l-k\}} \{T_j = T''_j\}\right].
$$

This proves our claim.

(4) *If $l \geq 2$ and $n_{l-k} = l$, then*

$$
P\left[\bigcap_{i \in \{1,\ldots,k\}} \{T_{m_i} = T'_i\} \cap \bigcap_{j \in \{1,\ldots,l-k\}} \{T_{n_j} = T''_j\}\right]
$$

$$
= \frac{\alpha''}{\alpha' + \alpha''} \cdot P\left[\bigcap_{i \in \{1,\ldots,k\}} \{T_{m_i} = T'_i\} \cap \bigcap_{j \in \{1,\ldots,l-k-1\}} \{T_{n_j} = T''_j\}\right].
$$

This means that elimination of $\{T_{n_{l-k}} = T''_l\}$ produces the factor $\alpha''/(\alpha' + \alpha'')$. The identity follows from (3) by symmetry.

(5) Using (1), (2), (3), and (4), the assertion now follows by induction over $l \in \mathbf{N}$. $\square$

It is clear that, for each $l \in \mathbf{N}$, the family $\{A_C\}_{C \in \mathcal{C}(l)}$ is disjoint; we shall now show that it is, up to a null set, even a partition of Ω.

6.4.4 Corollary. *The identity*

$$
\sum_{C \in \mathcal{C}(l)} P[A_C] = 1
$$

holds for all $l \in \mathbf{N}$.

Proof. By Lemma 6.4.3, we have

$$
\sum_{C \in \mathcal{C}(l)} P[A_C] = \sum_{k=0}^{l} \sum_{C \in \mathcal{C}(l,k)} P[A_C]
$$

$$
= \sum_{k=0}^{l} \binom{l}{k} \left(\frac{\alpha'}{\alpha' + \alpha''}\right)^k \left(\frac{\alpha''}{\alpha' + \alpha''}\right)^{l-k}
$$

$$
= 1,
$$

as was to be shown. $\square$

6.4.5 Corollary. *The identities*

$$\sum_{k=1}^{l} P[\{T_l = T_k'\}] \; = \; \frac{\alpha'}{\alpha' + \alpha''}$$

and

$$\sum_{k=1}^{l} P[\{T_l = T_k''\}] \; = \; \frac{\alpha''}{\alpha' + \alpha''}$$

hold for all $l \in \mathbf{N}$.

Proof. For $l \in \mathbf{N}$ and $k \in \{1, \ldots, l\}$, let $\mathcal{C}(l, k, k)$ denote the collection of all pairs $(\{m_i\}_{i \in \{1, \ldots, k\}}, \{n_j\}_{j \in \{1, \ldots, l-k\}}) \in \mathcal{C}(l)$ satisfying $m_k = l$. Then we have

$$\sum_{k=1}^{l} P[\{T_l = T_k'\}] \; = \; \sum_{k=1}^{l} \sum_{C \in \mathcal{C}(l,k,k)} P[A_C]$$

$$= \; \sum_{k=1}^{l} \sum_{C \in \mathcal{C}(l,k,k)} \left(\frac{\alpha'}{\alpha' + \alpha''}\right)^k \left(\frac{\alpha''}{\alpha' + \alpha''}\right)^{l-k}$$

$$= \; \sum_{k=1}^{l} \binom{l-1}{k-1} \left(\frac{\alpha'}{\alpha' + \alpha''}\right)^k \left(\frac{\alpha''}{\alpha' + \alpha''}\right)^{l-k}$$

$$= \; \frac{\alpha'}{\alpha' + \alpha''} \sum_{j=0}^{l-1} \binom{l-1}{j} \left(\frac{\alpha'}{\alpha' + \alpha''}\right)^j \left(\frac{\alpha''}{\alpha' + \alpha''}\right)^{(l-1)-j}$$

$$= \; \frac{\alpha'}{\alpha' + \alpha''} \,,$$

which is the first identity. The second identity follows by symmetry. □

For all $n \in \mathbf{N}$, define

$$X_n \; := \; \sum_{k=1}^{n} \left(\chi_{\{T_n = T_k'\}} X_k' + \chi_{\{T_n = T_k''\}} X_k''\right) .$$

The sequence $\{X_n\}_{n \in \mathbf{N}}$ is said to be the *superposition* of the claim size processes $\{X_n'\}_{n \in \mathbf{N}}$ and $\{X_n''\}_{n \in \mathbf{N}}$.

6.4.6 Theorem (Superposition of Claim Size Processes). *The sequence $\{X_n\}_{n \in \mathbf{N}}$ is i. i. d. and satisfies*

$$P_X \; = \; \frac{\alpha'}{\alpha' + \alpha''} P_{X'} + \frac{\alpha''}{\alpha' + \alpha''} P_{X''} .$$

Proof.   Consider $l \in \mathbf{N}$ and $B_1, \ldots, B_l \in \mathcal{B}(\mathbf{R})$.
For $C = (\{m_i\}_{i \in \{1,\ldots,k\}}, \{n_j\}_{j \in \{1,\ldots,l-k\}}) \in \mathcal{C}(l)$, Lemma 6.4.3 yields

$$P[A_C] = \left(\frac{\alpha'}{\alpha' + \alpha''}\right)^k \left(\frac{\alpha''}{\alpha' + \alpha''}\right)^{l-k} ,$$

and hence

$$P\left[\bigcap_{h=1}^{l} \{X_h \in B_h\} \cap A_C\right]$$

$$= P\left[\bigcap_{i \in \{1,\ldots,k\}} \{X'_{m_i} \in B_{m_i}\} \cap \bigcap_{j \in \{1,\ldots,l-k\}} \{X''_{n_j} \in B_{n_j}\} \cap A_C\right]$$

$$= \prod_{i \in \{1,\ldots,k\}} P[\{X'_{m_i} \in B_{m_i}\}] \cdot \prod_{j \in \{1,\ldots,l-k\}} P[\{X''_{n_j} \in B_{n_j}\}] \cdot P[A_C]$$

$$= \prod_{i \in \{1,\ldots,k\}} P_{X'}[B_{m_i}] \cdot \prod_{j \in \{1,\ldots,l-k\}} P_{X''}[B_{n_j}] \cdot \left(\frac{\alpha'}{\alpha' + \alpha''}\right)^k \left(\frac{\alpha''}{\alpha' + \alpha''}\right)^{l-k}$$

$$= \prod_{i \in \{1,\ldots,k\}} \frac{\alpha'}{\alpha' + \alpha''} P_{X'}[B_{m_i}] \cdot \prod_{j \in \{1,\ldots,l-k\}} \frac{\alpha''}{\alpha' + \alpha''} P_{X''}[B_{n_j}] .$$

By Corollary 6.4.4, summation over $\mathcal{C}(l)$ yields

$$P\left[\bigcap_{h=1}^{l} \{X_h \in B_h\}\right]$$

$$= \sum_{C \in \mathcal{C}(l)} P\left[\bigcap_{h=1}^{l} \{X_h \in B_h\} \cap A_C\right]$$

$$= \sum_{C \in \mathcal{C}(l)} \left(\prod_{i \in \{1,\ldots,k\}} \frac{\alpha'}{\alpha' + \alpha''} P_{X'}[B_{m_i}] \cdot \prod_{j \in \{1,\ldots,l-k\}} \frac{\alpha''}{\alpha' + \alpha''} P_{X''}[B_{n_j}]\right)$$

$$= \prod_{h=1}^{l} \left(\frac{\alpha'}{\alpha' + \alpha''} P_{X'}[B_h] + \frac{\alpha''}{\alpha' + \alpha''} P_{X''}[B_h]\right) ,$$

and the assertion follows. $\square$

We can now prove the main result of this section:

6.4.7 Theorem (Superposition of Poisson Risk Processes).   *The pair*
$(\{N_t\}_{t \in \mathbf{R}_+}, \{X_n\}_{n \in \mathbf{N}})$ *is a Poisson risk process.*

Proof.   Consider $l \in \mathbf{N}$, $B_1, \ldots, B_l \in \mathcal{B}(\mathbf{R})$, and a disjoint family $\{D_h\}_{h \in \{1,\ldots,l\}}$ of intervals in $(0, \infty)$ with increasing lower bounds. Define

$$\eta := \prod_{h=1}^{l-1} (\alpha' + \alpha'') \lambda[D_h] \cdot P_W[D_l] .$$

If we can show that

$$P\left[\bigcap_{h=1}^{l}\{X_h\in B_h\}\cap\{T_h\in D_h\}\right] \;=\; P\left[\bigcap_{h=1}^{l}\{X_h\in B_h\}\right]\cdot\eta\,,$$

then we have

$$P\left[\bigcap_{h=1}^{l}\{X_h\in B_h\}\cap\{T_h\in D_h\}\right] \;=\; P\left[\bigcap_{h=1}^{l}\{X_h\in B_h\}\right]\cdot P\left[\bigcap_{h=1}^{l}\{T_h\in D_h\}\right]\,.$$

We proceed in several steps:

(1) *The identity*

$$P\left[\bigcap_{h=1}^{l}\{T_h\in D_h\}\cap A_C\right] \;=\; \eta\cdot\left(\frac{\alpha'}{\alpha'+\alpha''}\right)^{k}\left(\frac{\alpha''}{\alpha'+\alpha''}\right)^{l-k}$$

holds for all $C=(\{m_i\}_{i\in\{1,\dots,k\}},\{n_j\}_{j\in\{1,\dots,l-k\}})\in\mathcal{C}(l)$.
Assume that $l=m_k$. For integration variables $v_1,\dots,v_k\in\mathbf{R}$ and $w_1,\dots,w_{l-k}\in\mathbf{R}$ and for $i\in\{0,1,\dots,k\}$ and $j\in\{0,1,\dots,l-k\}$, define

$$s_i \;:=\; \sum_{h=1}^{i} v_h$$

and

$$t_j \;:=\; \sum_{h=1}^{j} w_h\,.$$

Using translation invariance of the Lebesgue measure, we obtain

$$\int_{D_{n_1}-t_0}\cdots\int_{D_{n_{l-k}}-t_{l-k-1}} e^{\alpha''t_{l-k}}\,dP_{W_{l-k}''}(w_{l-k})\dots dP_{W_1''}(w_1) \;=\; \prod_{j=1}^{l-k}\alpha''\,\lambda[D_{n_j}]\,,$$

also, using the transformation formula, we obtain

$$\int_{D_{m_k}-s_{k-1}} e^{-\alpha''s_k}\,dP_{W_k'}(v_k)$$

$$=\; \int_{D_{m_k}-s_{k-1}} e^{-\alpha''s_k}\,\alpha'e^{-\alpha'v_k}\,\chi_{(0,\infty)}(v_k)\,d\lambda(v_k)$$

$$=\; \frac{\alpha'}{\alpha'+\alpha''}\,e^{\alpha's_{k-1}}\int_{D_{m_k}-s_{k-1}} (\alpha'+\alpha'')\,e^{-(\alpha'+\alpha'')(s_{k-1}+v_k)}\,\chi_{(0,\infty)}(v_k)\,d\lambda(v_k)$$

$$=\; \frac{\alpha'}{\alpha'+\alpha''}\,e^{\alpha's_{k-1}}\int_{D_{m_k}} (\alpha'+\alpha'')\,e^{-(\alpha'+\alpha'')s_k}\,\chi_{(0,\infty)}(s_k)\,d\lambda(s_k)$$

$$=\; \frac{\alpha'}{\alpha'+\alpha''}\,e^{\alpha's_{k-1}}\,P_W[D_{m_k}]$$

$$=\; \frac{\alpha'}{\alpha'+\alpha''}\,e^{\alpha's_{k-1}}\,P_W[D_l]\,,$$

and hence

$$\int_{D_{m_1}-s_0} \cdots \int_{D_{m_k}-s_{k-1}} e^{-\alpha'' s_k} dP_{W'_k}(v_k) \ldots dP_{W'_1}(v_1) = \frac{\alpha'}{\alpha'+\alpha''} P_W[D_l] \prod_{i=1}^{k-1} \alpha' \lambda[D_{m_i}] .$$

This yields

$$P\left[\bigcap_{h=1}^{l} \{T_h \in D_h\} \cap A_C\right]$$

$$= P\left[\bigcap_{i=1}^{k} \{T'_i \in D_{m_i}\} \cap \bigcap_{j=1}^{l-k} \{T''_j \in D_{n_j}\} \cap \{T'_k \le T''_{l-k+1}\}\right]$$

$$= P\left[\bigcap_{i=1}^{k} \{T'_i \in D_{m_i}\} \cap \bigcap_{j=1}^{l-k} \{T''_j \in D_{n_j}\} \cap \{T'_k \le T''_{l-k} + W''_{l-k+1}\}\right]$$

$$= \int_{D_{m_1}-s_0} \cdots \int_{D_{m_k}-s_{k-1}} \int_{D_{n_1}-t_0} \cdots \int_{D_{n_{l-k}}-t_{l-k-1}} \int_{(s_k-t_{l-k},\infty)} dP_{W''_{l-k+1}}(w_{l-k+1})$$
$$dP_{W''_{l-k}}(w_{l-k}) \ldots dP_{W''_1}(w_1) \, dP_{W'_k}(v_k) \ldots dP_{W'_1}(v_1)$$

$$= \int_{D_{m_1}-s_0} \cdots \int_{D_{m_k}-s_{k-1}} \int_{D_{n_1}-t_0} \cdots \int_{D_{n_{l-k}}-t_{l-k-1}} e^{-\alpha''(s_k-t_{l-k})}$$
$$dP_{W''_{l-k}}(w_{l-k}) \ldots dP_{W''_1}(w_1) \, dP_{W'_k}(v_k) \ldots dP_{W'_1}(v_1)$$

$$= \int_{D_{m_1}-s_0} \cdots \int_{D_{m_k}-s_{k-1}} e^{-\alpha'' s_k} \left(\int_{D_{n_1}-t_0} \cdots \int_{D_{n_{l-k}}-t_{l-k-1}} e^{\alpha'' t_{l-k}} \right.$$
$$\left. dP_{W''_{l-k}}(w_{l-k}) \ldots dP_{W''_1}(w_1) \right) dP_{W'_k}(v_k) \ldots dP_{W'_1}(v_1)$$

$$= \int_{D_{m_1}-s_0} \cdots \int_{D_{m_k}-s_{k-1}} e^{-\alpha'' s_k} \left(\prod_{j=1}^{l-k} \alpha'' \lambda[D_{n_j}] \right) dP_{W'_k}(v_k) \ldots dP_{W'_1}(v_1)$$

$$= \frac{\alpha'}{\alpha'+\alpha''} P_W[D_l] \prod_{i=1}^{k-1} \alpha' \lambda[D_{m_i}] \cdot \prod_{j=1}^{l-k} \alpha'' \lambda[D_{n_j}]$$

$$= \left(\frac{\alpha'}{\alpha'+\alpha''}\right)^k \left(\frac{\alpha''}{\alpha'+\alpha''}\right)^{l-k} \cdot \prod_{h=1}^{l-1} (\alpha'+\alpha'') \lambda[D_h] \cdot P_W[D_l]$$

$$= \left(\frac{\alpha'}{\alpha'+\alpha''}\right)^k \left(\frac{\alpha''}{\alpha'+\alpha''}\right)^{l-k} \cdot \eta .$$

By symmetry, the same result obtains in the case $l = n_{l-k}$.

(2) *The identity*

$$P\left[\bigcap_{h=1}^{l}\{X_h \in B_h\} \cap \{T_h \in D_h\} \cap A_C\right]$$

$$= \prod_{i=1}^{k} \frac{\alpha'}{\alpha' + \alpha''} P_{X'}[B_{m_i}] \cdot \prod_{j=1}^{l-k} \frac{\alpha''}{\alpha' + \alpha''} P_{X''}[B_{n_j}] \cdot \eta$$

holds for all $C = (\{m_i\}_{i\in\{1,\ldots,k\}}, \{n_j\}_{j\in\{1,\ldots,l-k\}}) \in \mathcal{C}(l)$.

Because of (1), we have

$$P\left[\bigcap_{h=1}^{l}\{X_h \in B_h\} \cap \{T_h \in D_h\} \cap A_C\right]$$

$$= P\left[\bigcap_{i=1}^{k}\{X'_{m_i} \in B_{m_i}\} \cap \{T'_{m_i} \in D_{m_i}\} \cap \bigcap_{j=1}^{l-k}\{X''_{n_j} \in B_{n_j}\} \cap \{T''_{n_j} \in D_{n_j}\} \cap A_C\right]$$

$$= P\left[\bigcap_{i=1}^{k}\{X'_{m_i} \in B_{m_i}\} \cap \bigcap_{j=1}^{l-k}\{X''_{n_j} \in B_{n_j}\}\right]$$

$$\cdot P\left[\bigcap_{i=1}^{k}\{T'_{m_i} \in D_{m_i}\} \cap \bigcap_{j=1}^{l-k}\{T''_{n_j} \in D_{n_j}\} \cap A_C\right]$$

$$= \prod_{i=1}^{k} P_{X'}[B_{m_i}] \prod_{j=1}^{l-k} P_{X''}[B_{n_j}] \cdot P\left[\bigcap_{h=1}^{l}\{T_h \in D_h\} \cap A_C\right]$$

$$= \prod_{i=1}^{k} P_{X'}[B_{m_i}] \prod_{j=1}^{l-k} P_{X''}[B_{n_j}] \cdot \left(\frac{\alpha'}{\alpha' + \alpha''}\right)^{k} \left(\frac{\alpha''}{\alpha' + \alpha''}\right)^{l-k} \eta$$

$$= \prod_{i=1}^{k} \frac{\alpha'}{\alpha' + \alpha''} P_{X'}[B_{m_i}] \cdot \prod_{j=1}^{l-k} \frac{\alpha''}{\alpha' + \alpha''} P_{X''}[B_{n_j}] \cdot \eta ,$$

as was to be shown.

(3) *We have*

$$P\left[\bigcap_{h=1}^{l}\{X_h \in B_h\} \cap \{T_h \in D_h\}\right] = P\left[\bigcap_{h=1}^{l}\{X_h \in B_h\}\right] \cdot \eta .$$

By Corollary 6.4.4 and because of (2) and Theorem 6.4.6, we have

$$P\left[\bigcap_{h=1}^{l}\{X_h \in B_h\} \cap \{T_h \in D_h\}\right]$$

$$= \sum_{C\in\mathcal{C}(l)} P\left[\bigcap_{h=1}^{l}\{X_h \in B_h\} \cap \{T_h \in D_h\} \cap A_C\right]$$

$$= \sum_{C \in \mathcal{C}(l)} \left(\prod_{i=1}^{k} \frac{\alpha'}{\alpha' + \alpha''} P_{X'}[B_{m_i}] \cdot \prod_{j=1}^{l-k} \frac{\alpha''}{\alpha' + \alpha''} P_{X''}[B_{n_j}] \cdot \eta \right)$$

$$= \eta \cdot \sum_{C \in \mathcal{C}(l)} \left(\prod_{i=1}^{k} \frac{\alpha'}{\alpha' + \alpha''} P_{X'}[B_{m_i}] \cdot \prod_{j=1}^{l-k} \frac{\alpha''}{\alpha' + \alpha''} P_{X''}[B_{n_j}] \right)$$

$$= \eta \cdot \prod_{h=1}^{l} \left(\frac{\alpha'}{\alpha' + \alpha''} P_{X'}[B_h] + \frac{\alpha''}{\alpha' + \alpha''} P_{X''}[B_h] \right)$$

$$= \eta \cdot \prod_{h=1}^{l} P[\{X_h \in B_h\}]$$

$$= \eta \cdot P\left[\bigcap_{h=1}^{l} \{X_h \in B_h\} \right],$$

as was to be shown.

(4) *We have*

$$P\left[\bigcap_{h=1}^{l} \{X_h \in B_h\} \cap \bigcap_{h=1}^{l} \{T_h \in D_h\} \right] = P\left[\bigcap_{h=1}^{l} \{X_h \in B_h\} \right] \cdot P\left[\bigcap_{h=1}^{l} \{T_h \in D_h\} \right].$$

This follows from (3).

(5) The previous identity remains valid if the sequence $\{D_h\}_{h \in \{1,\dots,l\}}$ of intervals is replaced by a sequence of general Borel sets. This implies that $\{X_n\}_{n \in \mathbb{N}}$ and $\{T_n\}_{n \in \mathbb{N}_0}$ are independent. □

Let us finally consider the aggregate claims process $\{S_t\}_{t \in \mathbb{R}_+}$ induced by the Poisson risk process $(\{N_t\}_{t \in \mathbb{R}_+}, \{X_n\}_{n \in \mathbb{N}})$. If the construction of the claim size process $\{X_n\}_{n \in \mathbb{N}}$ was appropriate, then the aggregate claims process $\{S_t\}_{t \in \mathbb{R}_+}$ should agree with the sum of the aggregate claims processes $\{S_t'\}_{t \in \mathbb{R}_+}$ and $\{S_t''\}_{t \in \mathbb{R}_+}$ induced by the original Poisson risk processes $(\{N_t'\}_{t \in \mathbb{R}_+}, \{X_n'\}_{n \in \mathbb{N}})$ and $(\{N_t''\}_{t \in \mathbb{R}_+}, \{X_n''\}_{n \in \mathbb{N}})$, respectively. The following result asserts that this is indeed true:

6.4.8 Theorem. *The identity*

$$S_t = S_t' + S_t''$$

holds for all $t \in \mathbb{R}_+$.

Proof. For all $\omega \in \{N_t = 0\}$, we clearly have

$$S_t(\omega) = S_t'(\omega) + S_t''(\omega).$$

Consider now $l \in \mathbb{N}$. For $C = (\{m_i\}_{i \in \{1,\dots,k\}}, \{n_j\}_{j \in \{1,\dots,l-k\}}) \in \mathcal{C}(l)$, we have

$$\{N_t = l\} \cap A_C = \{T_l \le t < T_{l+1}\} \cap A_C$$
$$= \{T_k' \le t < T_{k+1}'\} \cap \{T_{l-k}'' \le t < T_{l-k+1}''\} \cap A_C$$
$$= \{N_t' = k\} \cap \{N_t'' = l-k\} \cap A_C$$

and hence, for all $\omega \in \{N_t = l\} \cap A_C$,

$$
\begin{aligned}
S_t(\omega) &= \sum_{h=1}^{N_t(\omega)} X_h(\omega) \\
&= \sum_{h=1}^{l} X_h(\omega) \\
&= \sum_{i=1}^{k} X_{m_i}(\omega) + \sum_{j=1}^{l-k} X_{n_j}(\omega) \\
&= \sum_{i=1}^{N'_t(\omega)} X'_i(\omega) + \sum_{j=1}^{N''_t(\omega)} X''_j(\omega) \\
&= S'_t(\omega) + S''_t(\omega) \, .
\end{aligned}
$$

By Corollary 6.4.4, this yields

$$
S_t(\omega) = S'_t(\omega) + S''_t(\omega)
$$

for all $\omega \in \{N_t = l\}$.
We conclude that

$$
S_t = S'_t + S''_t \, ,
$$

as was to be shown.   □

Problems

6.4.A   Extend the results of this section to more than two independent Poisson risk processes.

6.4.B   Study the superposition problem for independent risk processes which are not Poisson risk processes.

6.4.C   **Discrete Time Model:** The sum of two independent binomial processes may fail to be a claim number process.

6.5   Remarks

For theoretical considerations, excess of loss reinsurance is probably the simplest form of reinsurance. Formally, excess of loss reinsurance with a priority held by the direct insurer is the same as direct insurance with a deductible to be paid by the insured.

For further information on reinsurance, see Gerathewohl [1976, 1979] and Dienst [1988]; for a discussion of the impact of deductibles, see Sterk [1979, 1980, 1988].

The superposition problem for renewal processes was studied by Störmer [1969].

Chapter 7

The Reserve Process and the Ruin Problem

In the present chapter we introduce the reserve process and study the ruin problem. We first extend the model considered so far and discuss some technical aspects of the ruin problem (Section 7.1). We next prove Kolmogorov's inequality for positive supermartingales (Section 7.2) and then apply this inequality to obtain Lundberg's inequality for the probability of ruin in the case where the excess premium process has a superadjustment coefficient (Section 7.3). We finally give some sufficient conditions for the existence of a superadjustment coefficient (Section 7.4).

7.1 The Model

Throughout this chapter, let $\{N_t\}_{t \in \mathbf{R}_+}$ be a claim number process, let $\{T_n\}_{n \in \mathbf{N}_0}$ be the claim arrival process induced by the claim number process, and let $\{W_n\}_{n \in \mathbf{N}}$ be the claim interarrival process induced by the claim arrival process. We assume that the exceptional null set is empty and that the probability of explosion is equal to zero.

Furthermore, let $\{X_n\}_{n \in \mathbf{N}}$ be a claim size process, let $\{S_t\}_{t \in \mathbf{R}_+}$ be the aggregate claims process induced by the claim number process and the claim size process, and let $\kappa \in (0, \infty)$. For $u \in (0, \infty)$ and all $t \in \mathbf{R}_+$, define

$$R_t^u := u + \kappa t - S_t .$$

Of course, we have $R_0^u = u$.

Interpretation:
– κ is the *premium intensity* so that κt is the *premium income* up to time t.
– u is the *initial reserve*.
– R_t^u is the *reserve* at time t when the initial reserve is u.
Accordingly, the family $\{R_t^u\}_{t \in \mathbf{R}_+}$ is said to be the *reserve process* induced by the

claim number process, the claim size process, the premium intensity, and the initial reserve.

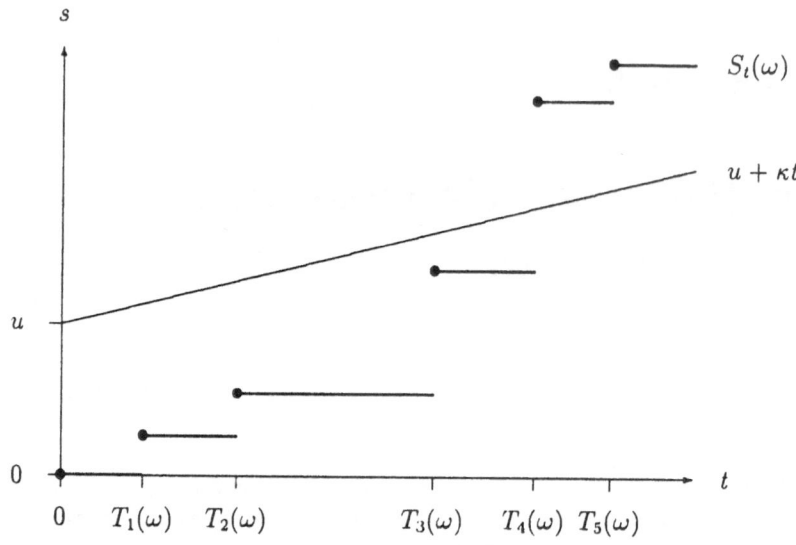

Claim Arrival Process and Aggregate Claims Process

We are interested in the *ruin problem* for the reserve process. This is the problem of calculating or estimating the probability of the event that the reserve process falls beyond zero at some time.

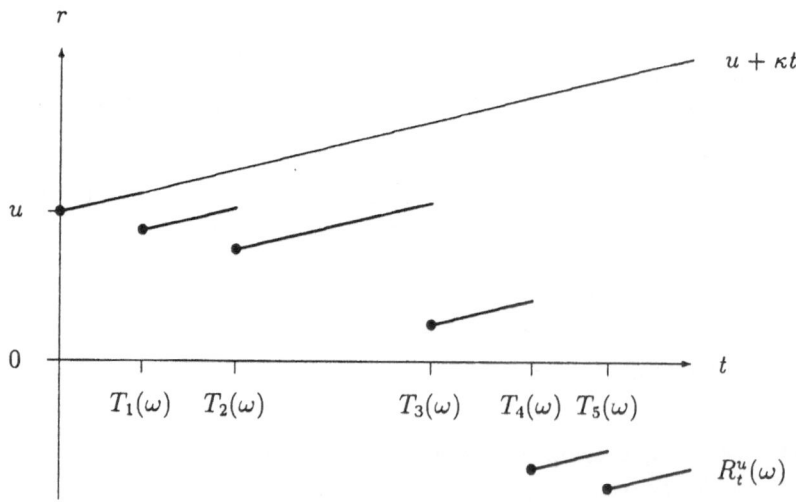

Claim Arrival Process and Reserve Process

In order to give a precise formulation of the ruin problem, we need the following measure–theoretic consideration:

Let $\{Z_t\}_{t\in\mathbf{R}_+}$ be family of measurable functions $\Omega \to [-\infty, \infty]$. A measurable function $Z : \Omega \to [-\infty, \infty]$ is said to be an *essential infimum* of $\{Z_t\}_{t\in\mathbf{R}_+}$ if it has the following properties:
- $P[\{Z \leq Z_t\}] = 1$ holds for all $t \in \mathbf{R}_+$, and
- every measurable function $Y : \Omega \to [-\infty, \infty]$ such that $P[\{Y \leq Z_t\}] = 1$ holds for all $t \in \mathbf{R}_+$ satisfies $P[\{Y \leq Z\}] = 1$.

The definition implies that any two essential infima of $\{Z_t\}_{t\in\mathbf{R}_+}$ are identical with probability one. The almost surely unique essential infimum of the family $\{Z_t\}_{t\in\mathbf{R}_+}$ is denoted by $\inf_{t\in\mathbf{R}_+} Z_t$.

7.1.1 Lemma.   *Every family $\{Z_t\}_{t\in\mathbf{R}_+}$ of measurable functions $\Omega \to [-\infty, \infty]$ possesses an essential infimum.*

Proof.   Without loss of generality, we may assume that $Z_t(\omega) \in [-1, 1]$ holds for all $t \in \mathbf{R}_+$ and $\omega \in \Omega$.

Let $\mathcal{J}$ denote the collection of all countable subsets of $\mathbf{R}_+$. For $J \in \mathcal{J}$, consider the measurable function Z_J satisfying

$$Z_J(\omega) \;=\; \inf_{t\in J} Z_t(\omega) \,.$$

Define

$$c \;:=\; \inf_{J\in\mathcal{J}} E[Z_J] \,,$$

choose a sequence $\{J_n\}_{n\in\mathbf{N}}$ satisfying

$$c \;=\; \inf_{n\in\mathbf{N}} E[Z_{J_n}] \,,$$

and define $J_\infty := \bigcup_{n\in\mathbf{N}} J_n$. Then we have $J_\infty \in \mathcal{J}$. Since $c \leq E[Z_{J_\infty}] \leq E[Z_{J_n}]$ holds for all $n \in \mathbf{N}$, we obtain

$$c \;=\; E[Z_{J_\infty}] \,.$$

For each $t \in \mathbf{R}_+$, we have $Z_{J_\infty \cup \{t\}} = Z_{J_\infty} \wedge Z_t$, hence

$$
\begin{aligned}
c &\leq& E[Z_{J_\infty\cup\{t\}}] \\
&=& E[Z_{J_\infty} \wedge Z_t] \\
&\leq& E[Z_{J_\infty}] \\
&=& c \,,
\end{aligned}
$$

whence

$$P[\{Z_{J_\infty} \wedge Z_t = Z_{J_\infty}\}] \;=\; 1 \,,$$

and thus

$$P[\{Z_{J_\infty} \leq Z_t\}] \;=\; 1 \,.$$

The assertion follows. $\square$

Let us now return to the reserve process $\{R_t^u\}_{t \in \mathbf{R}_+}$. By Lemma 7.1.1, the reserve process has an essential infimum which is denoted by

$$\inf_{t \in \mathbf{R}_+} R_t^u .$$

In particular, we have $\{\inf_{t \in \mathbf{R}_+} R_t^u < 0\} \in \mathcal{F}$. The event $\{\inf_{t \in \mathbf{R}_+} R_t^u < 0\}$ is called *ruin* of the reserve process and its probability is denoted by

$$\Psi(u) \quad := \quad P[\{\inf_{t \in \mathbf{R}_+} R_t^u < 0\}]$$

in order to emphasize the dependence of the probability of ruin on the initial reserve.

In practice, an upper bound Ψ^* for the probability of ruin is given in advance and the insurer is interested to choose the initial reserve u such that

$$\Psi(u) \quad \leq \quad \Psi^* .$$

In principle, one would like to choose u such that

$$\Psi(u) \quad = \quad \Psi^* ,$$

but the problem of computing the probability of ruin is even harder than the problem of computing the accumulated claims distribution. It is therefore desirable to have an upper bound $\Psi'(u)$ for the probability of ruin when the initial reserve is u, and to choose u such that

$$\Psi'(u) \quad = \quad \Psi^* .$$

Since

$$\Psi(u) \quad \leq \quad \Psi'(u) ,$$

the insurer is on the safe side but possibly binds too much capital.

It is intuitively clear that the time when the reserve process first falls beyond zero for the first time must be a claim arrival time. To make this point precise, we introduce a discretization of the reserve process: For $n \in \mathbf{N}$, define

$$G_n \quad := \quad \kappa W_n - X_n ,$$

and for $n \in \mathbf{N}_0$ define

$$U_n^u \quad := \quad u + \sum_{k=1}^n G_k .$$

Of course, we have $U_0^u = u$.

The sequence $\{G_n\}_{n \in \mathbf{N}}$ is said to be the *excess premium process*, and the sequence $\{U_n^u\}_{n \in \mathbf{N}_0}$ is said to be the *modified reserve process*.

The following result shows that the probability of ruin is determined by the modified reserve process:

7.1.2 Lemma. *The probability of ruin satisfies*

$$\Psi(u) = P[\{\inf_{n \in \mathbb{N}_0} U_n^u < 0\}] .$$

Proof. Define $A := \{\sup_{\mathbb{N}_0} T_n < \infty\}$. For each $\omega \in \Omega \backslash A$, we have

$$\inf_{t \in \mathbb{R}_+} R_t^u(\omega) = \inf_{t \in \mathbb{R}_+} \left(u + \kappa t - \sum_{k=1}^{N_t(\omega)} X_k(\omega) \right)$$

$$= \inf_{n \in \mathbb{N}_0} \inf_{t \in [T_n(\omega), T_{n+1}(\omega))} \left(u + \kappa t - \sum_{k=1}^{N_t(\omega)} X_k(\omega) \right)$$

$$= \inf_{n \in \mathbb{N}_0} \inf_{t \in [T_n(\omega), T_{n+1}(\omega))} \left(u + \kappa t - \sum_{k=1}^{n} X_k(\omega) \right)$$

$$= \inf_{n \in \mathbb{N}_0} \left(u + \kappa T_n(\omega) - \sum_{k=1}^{n} X_k(\omega) \right)$$

$$= \inf_{n \in \mathbb{N}_0} \left(u + \kappa \sum_{k=1}^{n} W_k(\omega) - \sum_{k=1}^{n} X_k(\omega) \right)$$

$$= \inf_{n \in \mathbb{N}_0} \left(u + \sum_{k=1}^{n} (\kappa W_k(\omega) - X_k(\omega)) \right)$$

$$= \inf_{n \in \mathbb{N}_0} \left(u + \sum_{k=1}^{n} G_k(\omega) \right)$$

$$= \inf_{n \in \mathbb{N}_0} U_n^u(\omega) .$$

Since the probability of explosion is assumed to be zero, we have $P[A] = 0$, hence

$$\inf_{t \in \mathbb{R}_+} R_t^u = \inf_{n \in \mathbb{N}_0} U_n^u ,$$

and thus

$$\Psi(u) = P[\{\inf_{t \in \mathbb{R}_+} R_t^u < 0\}]$$
$$= P[\{\inf_{n \in \mathbb{N}_0} U_n^u < 0\}] ,$$

as was to be shown. □

In the case where the excess premiums are i. i. d. with finite expectation, the previous
lemma yields a first result on the probability of ruin which sheds some light on the
different roles of the initial reserve and the premium intensity:

7.1.3 Theorem. *Assume that the sequence $\{G_n\}_{n \in \mathbb{N}}$ is i. i. d. with nondegenerate
distribution and finite expectation. If $E[G] \leq 0$, then, for every initial reserve, the
probability of ruin is equal to one.*

Proof. By the Chung–Fuchs Theorem, we have

$$
\begin{aligned}
1 &= P\left[\left\{ \liminf_{n \in \mathbb{N}} \sum_{k=1}^{n} G_k = -\infty \right\}\right] \\
&\leq P\left[\left\{ \liminf_{n \in \mathbb{N}} \sum_{k=1}^{n} G_k < -u \right\}\right] \\
&\leq P\left[\left\{ \inf_{n \in \mathbb{N}} \sum_{k=1}^{n} G_k < -u \right\}\right] \\
&= P[\{\inf_{n \in \mathbb{N}} U_n^u < 0\}] \,,
\end{aligned}
$$

and thus

$$
P[\{\inf_{n \in \mathbb{N}} U_n^u < 0\}] = 1 \,.
$$

The assertion now follows from Lemma 7.1.2.   □

7.1.4 Corollary. *Assume that the sequences $\{W_n\}_{n \in \mathbb{N}}$ and $\{X_n\}_{n \in \mathbb{N}}$ are independent and that each of them is i. i. d. with nondegenerate distribution and finite
expectation. If $\kappa \leq E[X]/E[W]$, then, for every initial reserve, the probability of
ruin is equal to one.*

It is interesting to note that Theorem 7.1.3 and Corollary 7.1.4 do not involve any
assumption on the initial reserve.

In the situation of Corollary 7.1.4 we see that, in order to prevent the probability of
ruin from being equal to one, the premium intensity must be large enough to ensure
that the expected premium income per claim, $\kappa E[W]$, is strictly greater than the
expected claim size $E[X]$. The expected claim size is called the *net premium*, and
the expected excess premium,

$$
E[G] = \kappa E[W] - E[X] \,,
$$

is said to be the *safety loading* of the (modified) reserve process. For later reference,
we note that the safety loading is strictly positive if and only if the premium intensity
safisfies

$$
\kappa > \frac{E[X]}{E[W]} \,.
$$

We shall return to the situation of Corollary 7.1.4 in Section 7.4 below.

7.2 Kolmogorov's Inequality for Positive Supermartingales

Our aim is to establish an upper bound for the probability of ruin under a suitable assumption on the excess premium process. The proof of this inequality will be based on Kolmogorov's inequality for positive supermartingales which is the subject of the present section.

Consider a sequence $\{Z_n\}_{n\in\mathbb{N}_0}$ of random variables having finite expectations and let $\{\mathcal{F}_n\}_{n\in\mathbb{N}_0}$ be the canonical filtration for $\{Z_n\}_{n\in\mathbb{N}_0}$.

A map $\tau : \Omega \to \mathbb{N}_0 \cup \{\infty\}$ is a *stopping time* for $\{\mathcal{F}_n\}_{n\in\mathbb{N}_0}$ if $\{\tau = n\} \in \mathcal{F}_n$ holds for all $n \in \mathbb{N}_0$, and it is *bounded* if $\sup_{\omega\in\Omega} \tau(\omega) < \infty$. Let $\mathbf{T}$ denote the collection of all bounded stopping times for $\{\mathcal{F}_n\}_{n\in\mathbb{N}_0}$, and note that $\mathbb{N}_0 \subseteq \mathbf{T}$.

For $\tau \in \mathbf{T}$, define

$$Z_\tau \; := \; \sum_{n=0}^{\infty} \chi_{\{\tau=n\}} Z_n \; .$$

Then Z_τ is a random variable satisfying

$$|Z_\tau| \; = \; \sum_{n=0}^{\infty} \chi_{\{\tau=n\}} |Z_n|$$

as well as

$$E[Z_\tau] \; = \; \sum_{n=0}^{\infty} \int_{\{\tau=n\}} Z_n \, dP \; .$$

Note that the sums occurring in the definition of Z_τ and in the formulas for $|Z_\tau|$ and $E[Z_\tau]$ actually extend only over a finite number of terms.

7.2.1 Lemma (Maximal Inequality). *The inequality*

$$P[\{\sup_{n\in\mathbb{N}_0} |Z_n| > \varepsilon\}] \; \leq \; \frac{1}{\varepsilon} \sup_{\tau\in\mathbf{T}} E|Z_\tau|$$

holds for all $\varepsilon \in (0, \infty)$.

Proof. Define

$$A_n \; := \; \{|Z_n| > \varepsilon\} \cap \bigcap_{k=1}^{n-1} \{|Z_k| \leq \varepsilon\} \; .$$

Then we have $\{\sup_{\mathbf{N}_0} |Z_n| > \varepsilon\} = \sum_{n=0}^{\infty} A_n$ and hence

$$P[\{\sup_{n \in \mathbf{N}_0} |Z_n| > \varepsilon\}] \;=\; \sum_{n=0}^{\infty} P[A_n] \,.$$

Consider $r \in \mathbf{N}_0$, and define a random variable τ_r by letting

$$\tau_r(\omega) \;:=\; \begin{cases} n & \text{if } \omega \in A_n \quad \text{and} \quad n \in \{0, 1, \ldots, r\} \\ r & \text{if } \omega \in \Omega \setminus \sum_{n=0}^{r} A_n \,. \end{cases}$$

Then we have $\tau_r \in \mathbf{T}$, and hence

$$
\begin{aligned}
\sum_{n=0}^{r} P[A_n] \;&\leq\; \sum_{n=0}^{r} \frac{1}{\varepsilon} \int_{A_n} |Z_n| \, dP \\
&=\; \frac{1}{\varepsilon} \sum_{n=0}^{r} \int_{A_n} |Z_{\tau_r}| \, dP \\
&\leq\; \frac{1}{\varepsilon} \int_{\Omega} |Z_{\tau_r}| \, dP \\
&\leq\; \frac{1}{\varepsilon} \sup_{\tau \in \mathbf{T}} E|Z_\tau| \,.
\end{aligned}
$$

Therefore, we have

$$
\begin{aligned}
P[\{\sup_{n \in \mathbf{N}_0} |Z_n| > \varepsilon\}] \;&=\; \sum_{n=0}^{\infty} P[A_n] \\
&\leq\; \frac{1}{\varepsilon} \sup_{\tau \in \mathbf{T}} E|Z_\tau| \,,
\end{aligned}
$$

as was to be shown. $\qquad\square$

7.2.2 Lemma.   *The following are equivalent:*
(a) $\{Z_n\}_{n \in \mathbf{N}_0}$ *is a supermartingale.*
(b) *The inequality* $E[Z_\sigma] \geq E[Z_\tau]$ *holds for all* $\sigma, \tau \in \mathbf{T}$ *such that* $\sigma \leq \tau$.

Proof.   Assume first that (a) holds and consider $\sigma, \tau \in \mathbf{T}$ such that $\sigma \leq \tau$. For all $k, n \in \mathbf{N}_0$ such that $n \geq k$, we have $\{\sigma = k\} \cap \{\tau \geq n{+}1\} = \{\sigma = k\} \setminus \{\tau \leq n\} \in \mathcal{F}_n$, and thus

$$
\begin{aligned}
\int_{\{\sigma=k\} \cap \{\tau \geq n\}} Z_n \, dP \;&=\; \int_{\{\sigma=k\} \cap \{\tau=n\}} Z_n \, dP + \int_{\{\sigma=k\} \cap \{\tau \geq n+1\}} Z_n \, dP \\
&\geq\; \int_{\{\sigma=k\} \cap \{\tau=n\}} Z_n \, dP + \int_{\{\sigma=k\} \cap \{\tau \geq n+1\}} Z_{n+1} \, dP \,.
\end{aligned}
$$

Induction yields

$$\int_{\{\sigma=k\}} Z_k \, dP = \int_{\{\sigma=k\}\cap\{\tau\geq k\}} Z_k \, dP$$

$$\geq \sum_{n=k}^{\infty} \int_{\{\sigma=k\}\cap\{\tau=n\}} Z_n \, dP \, ,$$

and this gives

$$\int_{\Omega} Z_\sigma \, dP = \sum_{k=0}^{\infty} \int_{\{\sigma=k\}} Z_k \, dP$$

$$\geq \sum_{k=0}^{\infty} \sum_{n=k}^{\infty} \int_{\{\sigma=k\}\cap\{\tau=n\}} Z_n \, dP$$

$$= \sum_{n=0}^{\infty} \sum_{k=0}^{n} \int_{\{\sigma=k\}\cap\{\tau=n\}} Z_n \, dP$$

$$= \sum_{n=0}^{\infty} \int_{\{\tau=n\}} Z_n \, dP$$

$$= \int_{\Omega} Z_\tau \, dP \, .$$

Therefore, (a) implies (b).

Assume now that (b) holds. Consider $n \in \mathbf{N}_0$ and $A \in \mathcal{F}_n$, and define a random variable τ by letting

$$\tau(\omega) := \begin{cases} n+1 & \text{if } \omega \in A \\ n & \text{if } \omega \in \Omega \backslash A \, . \end{cases}$$

Then we have $n \leq \tau \in \mathbf{T}$, hence

$$\int_A Z_n \, dP + \int_{\Omega\backslash A} Z_n \, dP = \int_{\Omega} Z_n \, dP$$

$$\geq \int_{\Omega} Z_\tau \, dP$$

$$= \int_A Z_{n+1} \, dP + \int_{\Omega\backslash A} Z_n \, dP \, ,$$

and thus

$$\int_A Z_n \, dP \geq \int_A Z_{n+1} \, dP \, .$$

Therefore, (b) implies (a). □

7.2.3 Lemma. *The following are equivalent:*
(a) $\{Z_n\}_{n\in\mathbf{N}_0}$ *is a martingale.*
(b) *The identity* $E[Z_\sigma] = E[Z_\tau]$ *holds for all* $\sigma, \tau \in \mathbf{T}$.

The proof of Lemma 7.2.3 is similar to that of Lemma 7.2.2.

The sequence $\{Z_n\}_{n\in\mathbf{N}_0}$ is *positive* if each Z_n is positive.

7.2.4 Corollary (Kolmogorov's Inequality). *If* $\{Z_n\}_{n\in\mathbf{N}_0}$ *is a positive supermartingale, then the inequality*

$$P[\{\sup_{n\in\mathbf{N}_0} Z_n > \varepsilon\}] \; \leq \; \frac{1}{\varepsilon} E[Z_0]$$

holds for all $\varepsilon \in (0, \infty)$.

This is immediate from Lemmas 7.2.1 and 7.2.2.

7.3   Lundberg's Inequality

Throughout this section, we assume that the sequence of excess premiums $\{G_n\}_{n\in\mathbf{N}}$ is independent.

A constant $\varrho \in (0, \infty)$ is a *superadjustment coefficient* for the excess premium process $\{G_n\}_{n\in\mathbf{N}}$ if it satisfies

$$E\left[e^{-\varrho G_n}\right] \; \leq \; 1$$

for all $n \in \mathbf{N}$, and it is an *adjustment coefficient* for the excess premium process if it satisfies

$$E\left[e^{-\varrho G_n}\right] \; = \; 1$$

for all $n \in \mathbf{N}$. The excess premium process need not possess a superadjustment coefficient; if the distribution of some excess premium is nondegenerate, then the excess premium process has at most one adjustment coefficient.

Let $\{\mathcal{F}_n\}_{n\in\mathbf{N}_0}$ denote the canonical filtration for $\{U_n^u\}_{n\in\mathbf{N}_0}$.

7.3.1 Lemma. *For* $\varrho \in (0, \infty)$, *the identity*

$$\int_A e^{-\varrho U_{n+1}^u} \, dP \; = \; \int_A e^{-\varrho U_n^u} \, dP \cdot \int_\Omega e^{-\varrho G_{n+1}} \, dP$$

holds for all $n \in \mathbf{N}_0$ *and* $A \in \mathcal{F}_n$.

Proof. For all $n \in \mathbf{N}_0$, we have

$$\begin{aligned} \mathcal{F}_n &= \sigma(\{U_k^u\}_{k \in \{0,1,...,n\}}) \\ &= \sigma(\{G_k\}_{k \in \{1,...,n\}}) \ . \end{aligned}$$

Since the sequence $\{G_n\}_{n \in \mathbf{N}}$ is independent, this yields

$$\begin{aligned} \int_A e^{-\varrho U_{n+1}^u} \, dP &= \int_\Omega \chi_A e^{-\varrho(U_n^u + G_{n+1})} \, dP \\ &= \int_\Omega \chi_A e^{-\varrho U_n^u} e^{-\varrho G_{n+1}} \, dP \\ &= \int_\Omega \chi_A e^{-\varrho U_n^u} \, dP \cdot \int_\Omega e^{-\varrho G_{n+1}} \, dP \\ &= \int_A e^{-\varrho U_n^u} \, dP \cdot \int_\Omega e^{-\varrho G_{n+1}} \, dP \ , \end{aligned}$$

for all $n \in \mathbf{N}_0$ and $A \in \mathcal{F}_n$. The assertion follows. $\qquad\square$

As an immediate consequence of Lemma 7.3.1, we obtain the following characterizations of superadjustment coefficients and adjustment coefficients:

7.3.2 Corollary. *For $\varrho \in (0, \infty)$, the following are equivalent:*
(a) *ϱ is a superadjustment coefficient for the excess premium process.*
(b) *For every $u \in (0, \infty)$, the sequence $\{e^{-\varrho U_n^u}\}_{n \in \mathbf{N}_0}$ is a supermartingale.*

7.3.3 Corollary. *For $\varrho \in (0, \infty)$, the following are equivalent:*
(a) *ϱ is an adjustment coefficient for the excess premium process.*
(b) *For every $u \in (0, \infty)$, the sequence $\{e^{-\varrho U_n^u}\}_{n \in \mathbf{N}_0}$ is a martingale.*

The main result of this section is the following:

7.3.4 Theorem (Lundberg's Inequality). *If $\varrho \in (0, \infty)$ is a superadjustment coefficient for the excess premium process, then the identity*

$$P[\{\inf_{n \in \mathbf{N}_0} U_n^u < 0\}] \leq e^{-\varrho u}$$

holds for all $u \in (0, \infty)$.

Proof. By Corollaries 7.3.2 and 7.2.4, we have

$$\begin{aligned} P[\{\inf_{n \in \mathbf{N}_0} U_n^u < 0\}] &= P[\{\sup_{n \in \mathbf{N}_0} e^{-\varrho U_n^u} > 1\}] \\ &\leq E[e^{-\varrho U_0^u}] \\ &= E[e^{-\varrho u}] \\ &= e^{-\varrho u} \ , \end{aligned}$$

as was to be shown. $\qquad\square$

The upper bound for the probability of ruin provided by Lundberg's inequality depends explicitly on the initial reserve u. Implicitly, it also depends, via the superadjustment coefficient ϱ, on the premium intensity κ.

Problem

7.3.A   Assume that the sequence $\{X_n\}_{n\in\mathbf{N}}$ is independent. If there exists some $\varrho \in (0,\infty)$ satisfying $E[e^{\varrho X_n}] \leq 1$ for all $n \in \mathbf{N}$, then the inequality

$$P[\sup_{t\in\mathbf{R}_+} S_t > c] \ \leq \ e^{-\varrho c}$$

holds for all $c \in (0,\infty)$.

Hint: Extend Lemma 7.1.2 and Lundberg's inequality to the case $\kappa = 0$.

7.4   On the Existence of a Superadjustment Coefficient

In the present section, we study the existence of a (super)adjustment coefficient.

We first consider the case where the excess premiums are i. i. d. According to the following result, we have to assume that the safety loading is strictly positive:

7.4.1 Theorem.   *Assume that the sequence $\{G_n\}_{n\in\mathbf{N}}$ is i. i. d. with nondegenerate distribution and finite expectation. If the excess premium process has a superadjustment coefficient, then $E[G] > 0$.*

Proof.   The assertion follows from Theorems 7.3.4 and 7.1.3.   □

7.4.2 Corollary.   *Assume that the sequences $\{W_n\}_{n\in\mathbf{N}}$ and $\{X_n\}_{n\in\mathbf{N}}$ are independent and that each of them is i. i. d. with nondegenerate distribution and finite expectation. If the excess premium process has a superadjustment coefficient, then $\kappa > E[X]/E[W]$.*

The previous result has a partial converse:

7.4.3 Theorem.   *Assume that*
(i) $\{W_n\}_{n\in\mathbf{N}}$ and $\{X_n\}_{n\in\mathbf{N}}$ are independent,
(ii) $\{W_n\}_{n\in\mathbf{N}}$ is i. i. d. and satisfies $\sup\{z \in \mathbf{R}_+ \mid E[e^{zW}] < \infty\} \in (0,\infty]$, and
(iii) $\{X_n\}_{n\in\mathbf{N}}$ is i. i. d. and satisfies $\sup\{z \in \mathbf{R}_+ \mid E[e^{zX}] < \infty\} \in (0,\infty)$ as well as $P_X[\mathbf{R}_+] = 1$.
If $\kappa > E[X]/E[W]$, then the excess premium process has an adjustment coefficient.

Proof.   By assumption, the sequence $\{G_n\}_{n\in\mathbf{N}}$ is i. i. d. For all $z \in \mathbf{R}$, we have

$$E\big[e^{-zG}\big] \ = \ E\big[e^{-z\kappa W}\big]\, E\big[e^{zX}\big]\,,$$

and hence

$$M_G(-z) \ = \ M_W(-\kappa z)\, M_X(z)\,.$$

By assumption, there exists some $z' \in (0, \infty)$ such that the moment generating functions of W and of X are both finite on the interval $(-\infty, z')$. Differentiation gives

$$-M_G'(z) \;=\; -\kappa \, M_W'(-\kappa z) \, M_X(z) + M_W(-\kappa z) \, M_X'(z)$$

for all z in a neighbourhood of 0, and thus

$$M_G'(0) \;=\; \kappa \, E[W] - E[X] \,.$$

By assumption, there exists some $z^* \in (0, \infty)$ satisfying $M_X(z^*) = \infty$ and hence $M_G(-z^*) = \infty$. Since $M_G(0) = 1$ and $M_G'(0) > 0$, it follows that there exists some $\varrho \in (0, z^*)$ satisfying $E[e^{-\varrho G}] = M_G(-\varrho) = 1$. But this means that ϱ is an adjustment coefficient for the excess premium process. $\qquad\square$

In Corollary 7.4.2 and Theorem 7.4.3, the claim interarrival times are i.i.d., which means that the claim number process is a renewal process. A particular renewal process is the Poisson process:

7.4.4 Corollary. *Assume that*
(i) *$\{N_t\}_{t \in \mathbf{R}_+}$ and $\{X_n\}_{n \in \mathbf{N}}$ are independent,*
(ii) *$\{N_t\}_{t \in \mathbf{R}_+}$ is a Poisson process with parameter α, and*
(iii) *$\{X_n\}_{n \in \mathbf{N}}$ is i.i.d. and satisfies $\sup\{z \in \mathbf{R}_+ \mid E[e^{zX}] < \infty\} \in (0, \infty)$ as well as $P_X[\mathbf{R}_+] = 1$.*
If $\kappa > \alpha E[X]$, then the excess premium process has an adjustment coefficient.

Proof. By Lemmas 2.1.3 and 1.1.1, the claim interarrival process $\{W_n\}_{n \in \mathbf{N}}$ and the claim size process $\{X_n\}_{n \in \mathbf{N}}$ are independent.
By Theorem 2.3.4, the sequence $\{W_n\}_{n \in \mathbf{N}}$ is i.i.d. with $P_W = \mathbf{Exp}(\alpha)$, and this yields $\sup\{z \in \mathbf{R}_+ \mid E[e^{zW}] < \infty\} = \alpha$ and $E[W] = 1/\alpha$.
The assertion now follows from Theorem 7.4.3. $\qquad\square$

In order to apply Lundberg's inequality, results on the existence of an adjustment coefficient are, of course, not sufficient; instead, the adjustment coefficient has to be determined explicitly. To this end, the distributions of the excess premiums have to be specified, and this is usually done by specifying the distributions of the claim interarrival times and those of the claim severities.

7.4.5 Theorem. *Assume that*
(i) *$\{N_t\}_{t \in \mathbf{R}_+}$ and $\{X_n\}_{n \in \mathbf{N}}$ are independent,*
(ii) *$\{N_t\}_{t \in \mathbf{R}_+}$ is a Poisson process with parameter α, and*
(iii) *$\{X_n\}_{n \in \mathbf{N}}$ is i.i.d. and satisfies $P_X = \mathbf{Exp}(\beta)$.*
If $\kappa > \alpha/\beta$, then $\beta - \alpha/\kappa$ is an adjustment coefficient for the excess premium process. In particular,

$$P[\{\inf_{n \in \mathbf{N}} U_n^u < 0\}] \;\leq\; e^{-(\beta - \alpha/\kappa)u} \,.$$

Proof.   By Theorem 2.3.4, the sequence $\{W_n\}_{n\in\mathbb{N}}$ is i.i.d. with $P_W = \mathbf{Exp}(\alpha)$. Define

$$\varrho := \beta - \frac{\alpha}{\kappa}.$$

Then we have

$$
\begin{aligned}
E\big[e^{-\varrho G}\big] &= E\big[e^{-\varrho\kappa W}\big]\, E\big[e^{\varrho X}\big] \\
&= M_W(-\varrho\kappa)\, M_X(\varrho) \\
&= \frac{\alpha}{\alpha + \varrho\kappa}\, \frac{\beta}{\beta - \varrho} \\
&= 1\,,
\end{aligned}
$$

which means that ϱ is an adjustment coefficient for the excess premium process. The inequality for the probability of ruin follows from Theorem 7.3.4.   □

In the previous result, the bound for the probability of ruin decreases when either the initial reserve or the premium intensity increases.

Let us finally turn to a more general situation in which the excess premiums are still independent but need not be identically distributed. In this situation, the existence of an adjustment coefficient cannot be expected, and superadjustment coefficients come into their own right:

7.4.6 Theorem.   *Let $\{\alpha_n\}_{n\in\mathbb{N}}$ and $\{\beta_n\}_{n\in\mathbb{N}}$ be two sequences of real numbers in $(0,\infty)$ such that $\alpha := \sup_{n\in\mathbb{N}} \alpha_n < \infty$ and $\beta := \inf_{n\in\mathbb{N}} \beta_n > 0$. Assume that*
(i) $\{N_t\}_{t\in\mathbb{R}_+}$ and $\{X_n\}_{n\in\mathbb{N}}$ are independent,
(ii) $\{N_t\}_{t\in\mathbb{R}_+}$ is a regular Markov process with intensities $\{\lambda_n\}_{n\in\mathbb{N}}$ satisfying $\lambda_n(t) = \alpha_n$ for all $n \in \mathbb{N}$ and $t \in \mathbb{R}_+$, and
(iii) $\{X_n\}_{n\in\mathbb{N}}$ is independent and satisfies $P_{X_n} = \mathbf{Exp}(\beta_n)$ for all $n \in \mathbb{N}$.
If $\kappa > \alpha/\beta$, then $\beta - \alpha/\kappa$ is a superadjustment coefficient for the excess premium process. In particular,

$$P[\{\inf_{n\in\mathbb{N}} U_n^u < 0\}] \leq e^{-(\beta - \alpha/\kappa)u}.$$

Proof.   By Theorem 3.4.2, the sequence $\{W_n\}_{n\in\mathbb{N}}$ is independent and satisfies $P_{W_n} = \mathbf{Exp}(\alpha_n)$ for all $n \in \mathbb{N}$. Define

$$\varrho := \beta - \frac{\alpha}{\kappa}.$$

As in the proof of Theorem 7.4.5, we obtain

$$E\big[e^{-\varrho G_n}\big] = E\big[e^{-\varrho\kappa W_n}\big]\, E\big[e^{\varrho X_n}\big]$$

$$= E\left[e^{-\varrho\kappa W}\right] E[e^{\varrho X}]$$

$$= \frac{\alpha_n}{\alpha_n + \varrho\kappa} \frac{\beta_n}{\beta_n - \varrho}$$

$$= \frac{\alpha_n\beta_n}{\alpha_n\beta_n + \varrho(\beta_n\kappa - \varrho\kappa - \alpha_n)}$$

$$= \frac{\alpha_n\beta_n}{\alpha_n\beta_n + \varrho((\beta_n - \beta)\kappa + (\alpha - \alpha_n))}$$

$$\leq 1,$$

which means that ϱ is a superadjustment coefficient for the excess premium process. The inequality for the probability of ruin follows from Theorem 7.3.4. $\square$

The previous result, which includes Theorem 7.4.5 as a special case, is a rather exceptional example where a superadjustment coefficient does not only exist but can even be given in explicit form.

Problems

7.4.A   In Theorem 7.4.3 and Corollary 7.4.4, the condition on M_X is fulfilled whenever $P_X = \mathbf{Exp}(\beta)$, $P_X = \mathbf{Geo}(\eta)$, or $P_X = \mathbf{Log}(\eta)$.

7.4.B   **Discrete Time Model:** Assume that
   (i) $\{N_l\}_{l \in \mathbf{N}_0}$ and $\{X_n\}_{n \in \mathbf{N}}$ are independent,
   (ii) $\{N_l\}_{l \in \mathbf{N}_0}$ is a binomial process with parameter ϑ, and
   (iii) $\{X_n\}_{n \in \mathbf{N}}$ is i.i.d. and satisfies $\sup\{z \in \mathbf{R}_+ \mid E[e^{zX}] < \infty\} \in (0, \infty)$ as well
         as $P_X[\mathbf{R}_+] = 1$.
   If $\kappa > \vartheta E[X]$, then the excess premium process has an adjustment coefficient.

7.5 Remarks

In the discussion of the ruin problem, we have only considered a fixed premium intensity and a variable initial reserve. We have done so in order not to overburden the notation and to clarify the role of (super)adjustment coefficients, which depend on the premium intensity but not on the initial reserve. Of course, the premium intensitiy may be a decision variable as well which can be determined by a given upper bound for the probability of ruin, but the role of the initial reserve and the role of the premium intensity are nevertheless quite different since the former is limited only by the financial power of the insurance company while the latter is to a large extent constrained by market conditions.

The (super)martingale approach to the ruin problem is due to Gerber [1973, 1979] and has become a famous method in ruin theory; see also DeVylder [1977], Delbaen and Haezendonck [1985], Rhiel [1986, 1987], Björk and Grandell [1988], Dassios and

Embrechts [1989], Grandell [1991], Møller [1992], Embrechts, Grandell, and Schmidli [1993], Embrechts and Schmidli [1994], Møller [1995], and Schmidli [1995].

The proof of Kolmogorov's inequality is usually based on the nontrivial fact that a supermartingale $\{Z_n\}_{n \in \mathbb{N}_0}$ satisfies $E[Z_0] \geq E[Z_\tau]$ for arbitrary stopping times τ; see Neveu [1972]. The simple proof presented here is well–known in the theory of asymptotic martingales; see Gut and Schmidt [1983] for a survey and references.

Traditionally, Lundberg's inequality is proven under the assumption that ϱ is an adjustment coefficient; the extension to the case of a superadjustment coefficient is due to Schmidt [1989]. The origin of this extension, which is quite natural with regard to the use of Kolmogorov's inequality and the relation between (super)adjustment coefficients and (super)martingales, is in a paper by Mammitzsch [1986], who also pointed out that in the case of i. i. d. excess premiums a superadjustment coefficient may exist when an adjustment does not exist.

For a discussion of the estimation problem for the (super)adjustment coefficient, see Herkenrath [1986], Deheuvels and Steinebach [1990], Csörgö and Steinebach [1991], Embrechts and Mikosch [1991], and Steinebach [1993].

Although Theorem 7.4.3 provides a rather general condition under which an adjustment coefficient exists, there are important claim size distributions which do not satisfy these conditions; an example is the Pareto distribution, which assigns high probability to large claims. For a discussion of the ruin problem for such *heavy tailed* claim size distributions, see Thorin and Wikstad [1977], Seal [1980], Embrechts and Veraverbeke [1982], Embrechts and Villaseñor [1988], Klüppelberg [1989], and Beirlant and Teugels [1992].

A natural extension of the model considered in this chapter is to assume that the premium income is not deterministic but stochastic; see Bühlmann [1972], DeVylder [1977], Dassios and Embrechts [1989], Dickson [1991], and Møller [1992].

While the (homogeneous) Poisson process still plays a prominent role in ruin theory, there are two major classes of claim number processes, renewal processes and Cox processes, which present quite different extensions of the Poisson process and for which the probability of ruin has been studied in detail. Recent work focusses on *Cox processes*, or *doubly stochastic Poisson processes*, which are particularly interesting since they present a common generalization of the inhomogeneous Poisson process and the mixed Poisson process; see e. g. Grandell [1991] and the references given there.

Let us finally remark that several authors have also studied the probability that the reserve process attains negative values in a bounded time interval; for a discussion of such *finite time ruin probabilities*, see again Grandell [1991].

Appendix: Special Distributions

In this appendix we recall the definitions and some properties of the probability distributions which are used or mentioned in the text. For comments on applications of these and other distributions in risk theory, see Panjer and Willmot [1992].

Auxiliary Notions

The Gamma Function

The map $\Gamma : (0, \infty) \to (0, \infty)$ given by

$$\Gamma(\gamma) \; := \; \int_0^\infty e^{-x}\, x^{\gamma-1}\, dx$$

is called the *gamma function*. It has the following properties:

$$\begin{aligned}
\Gamma(1/2) &= \sqrt{\pi} \\
\Gamma(1) &= 1 \\
\Gamma(\gamma+1) &= \gamma\,\Gamma(\gamma)
\end{aligned}$$

In particular, the identity

$$\Gamma(n+1) \; = \; n!$$

holds for all $n \in \mathbf{N}_0$. Roughly speaking, the values of the gamma function correspond to factorials.

The Beta Function

The map $B : (0, \infty) \times (0, \infty)$ given by

$$B(\alpha, \beta) \; := \; \int_0^1 x^{\alpha-1}\,(1-x)^{\beta-1}\, dx$$

is called the *beta function*. The fundamental identity for the beta function is

$$B(\alpha, \beta) \; = \; \frac{\Gamma(\alpha)\,\Gamma(\beta)}{\Gamma(\alpha+\beta)} \; ,$$

showing that the properties of the beta function follow from those of the gamma function. Roughly speaking, the inverted values of the beta function correspond to binomial coefficients.

The Generalized Binomial Coefficient

For $\alpha \in \mathbf{R}$ and $m \in \mathbf{N}_0$, the *generalized binomial coefficient* is defined to be

$$\binom{\alpha}{m} := \prod_{j=0}^{m-1} \frac{\alpha - j}{m - j} .$$

For $\alpha \in (0, \infty)$, the properties of the gamma function yield the identity

$$\binom{\alpha + m - 1}{m} = \frac{\Gamma(\alpha + m)}{\Gamma(\alpha) \, m!}$$

which is particularly useful.

Measures

We denote by $\boldsymbol{\xi} : \mathcal{B}(\mathbf{R}) \to \mathbf{R}$ the *counting measure* concentrated on $\mathbf{N}_0$, and we denote by $\boldsymbol{\lambda} : \mathcal{B}(\mathbf{R}) \to \mathbf{R}$ the *Lebesgue measure*. These measures are σ-finite, and the most important probability measures $\mathcal{B}(\mathbf{R}) \to [0,1]$ are absolutely continuous wiht respect to either $\boldsymbol{\xi}$ or $\boldsymbol{\lambda}$.

For $n \in \mathbf{N}$, we denote by $\boldsymbol{\lambda}^n : \mathcal{B}(\mathbf{R}^n) \to \mathbf{R}$ the n-dimensional *Lebesgue measure*.

Generalities on Distributions

A probability measure $Q : \mathcal{B}(\mathbf{R}^n) \to [0,1]$ is called a *distribution*.

A distribution Q is *degenerate* if there exists some $y \in \mathbf{R}^n$ satisfying

$$Q[\{y\}] = 1 ,$$

and it is *nondegenerate* if it is not degenerate.

In the remainder of this appendix, we consider only distributions $\mathcal{B}(\mathbf{R}) \to [0,1]$.

For $y \in \mathbf{R}$, the *Dirac distribution* $\boldsymbol{\delta}_y$ is defined to be the (degenerate) distribution Q satisfying

$$Q[\{y\}] = 1 .$$

Because of the particular role of the Dirac distribution, all parametric classes of distributions considered below are defined as to exclude degenerate distributions.

Let Q and R be distributions $\mathcal{B}(\mathbf{R}) \to [0,1]$.

Expectation and Higher Moments

If

$$\min \left\{ \int_{(-\infty,0]} (-x) \, dQ(x), \int_{[0,\infty)} x \, dQ(x) \right\} \ < \ \infty \,,$$

then the *expectation* of Q is said to *exist* and is defined to be

$$E[Q] \ := \ \int_{\mathbf{R}} x \, dQ(x) \,;$$

if

$$\max \left\{ \int_{(-\infty,0]} (-x) \, dQ(x), \int_{[0,\infty)} x \, dQ(x) \right\} \ < \ \infty$$

or, equivalently,

$$\int_{\mathbf{R}} |x| \, dQ(x) \ < \ \infty \,,$$

then the expectation of Q exists and is said to be *finite*. In this case, Q is said to have *finite expectation*.

More generally, if, for some $n \in \mathbf{N}$,

$$\int_{\mathbf{R}} |x|^n \, dQ(x) \ < \ \infty \,,$$

then Q is said to have a *finite moment of order n* or to have a *finite n-th moment* and the n-th moment of Q is defined to be

$$\int_{\mathbf{R}} x^n \, dQ(x) \,.$$

If Q has a finite moment of order n, then it also has a finite moment of order k for all $k \in \{1, \ldots, n-1\}$. The distribution Q is said to have *finite moments of any order* if

$$\int_{\mathbf{R}} |x|^n \, dQ(x) \ < \ \infty$$

holds for all $n \in \mathbf{N}$.

Variance and Coefficient of Variation

If Q has finite expectation, then the *variance* of Q is defined to be

$$\mathrm{var}\,[Q] \ := \ \int_{\mathbf{R}} (x - E[Q])^2 \, dQ(x) \,.$$

If Q satisfies $Q[\mathbf{R}_+] = 1$ and $E[Q] \in (0, \infty)$, then the *coefficient of variation* of Q is defined to be

$$v[Q] \ := \ \frac{\sqrt{\mathrm{var}\,[Q]}}{E[Q]} \,.$$

Characteristic Function

The *characteristic function* or *Fourier transform* of Q is defined to be the map $\varphi_Q : \mathbf{R} \to \mathbf{C}$ given by

$$\varphi_Q(z) \ := \ \int_{\mathbf{R}} e^{izx} \, dQ(x) \, .$$

Obviously, $\varphi_Q(0) = 1$. Moreover, a deep result on Fourier transforms asserts that the distribution Q is uniquely determined by its characteristic function φ_Q.

Moment Generating Function

The *moment generating function* of Q is defined to be the map $M_Q : \mathbf{R} \to [0, \infty]$ given by

$$M_Q(z) \ := \ \int_{\mathbf{R}} e^{zx} \, dQ(x) \, .$$

Again, $M_Q(0) = 1$. Moreover, if the moment generating function of Q is finite in a neighbourhood of zero, then Q has finite moments of any order and the identity

$$\frac{d^n M_Q}{dz^n}(0) \ = \ \int_{\mathbf{R}} x^n \, dQ(x)$$

holds for all $n \in \mathbf{N}$.

Probability Generating Function

If $Q[\mathbf{N}_0] = 1$, then the *probability generating function* of Q is defined to be the map $m_Q : [-1, 1] \to \mathbf{R}$ given by

$$m_Q(z) \ := \ \int_{\mathbf{R}} z^x \, dQ(x)$$
$$= \ \sum_{n=0}^{\infty} z^n \, Q[\{n\}] \, .$$

Since the identity

$$\frac{1}{n!} \frac{d^n m_Q}{dz^n}(0) \ = \ Q[\{n\}]$$

holds for all $n \in \mathbf{N}_0$, the distribution Q is uniquely determined by its probability generating function m_Q. The probability generating function has a unique extension to the closed unit disc in the complex plane.

Convolution

If $+ : \mathbf{R}^2 \to \mathbf{R}$ is defined to be the map given by $+(x, y) := x + y$, then

$$Q * R := (Q \otimes R)_+$$

is a distribution which is called the *convolution* of Q and R. The convolution satisfies

$$\varphi_{Q*R} = \varphi_Q \cdot \varphi_R \,,$$

and hence $Q * R = R * Q$, as well as

$$M_{Q*R} = M_Q \cdot M_R \,;$$

also, if $Q[\mathbf{N}_0] = 1 = R[\mathbf{N}_0]$, then

$$m_{Q*R} = m_Q \cdot m_R \,.$$

If Q and R have finite expectations, then

$$E[Q * R] = E[Q] + E[R] \,,$$

and if Q and R both have a finite second moment, then

$$\text{var}\,[Q * R] = \text{var}\,[Q] + \text{var}\,[R] \,.$$

Furthermore, the identity

$$(Q * R)[B] = \int_{\mathbf{R}} Q[B - y] \, dR(y)$$

holds for all $B \in \mathcal{B}(\mathbf{R})$; in particular, $Q * \boldsymbol{\delta}_y = \boldsymbol{\delta}_y * Q$ is the translation of Q by y. If $Q = \int f \, d\nu$ and $R = \int g \, d\nu$ for $\nu \in \{\boldsymbol{\xi}, \boldsymbol{\lambda}\}$, then $Q * R = \int f*g \, d\nu$, where the map $f*g : \mathbf{R} \to \mathbf{R}_+$ is defined by

$$(f*g)(x) := \int_{\mathbf{R}} f(x - y) g(y) \, d\nu(y) \,.$$

For $n \in \mathbf{N}_0$, the n-fold convolution of Q is defined to be

$$Q^{*n} := \begin{cases} \boldsymbol{\delta}_0 & \text{if } n = 0 \\ Q * Q^{*(n-1)} & \text{if } n \in \mathbf{N} \end{cases}$$

If $Q = \int f \, d\nu$ for $\nu \in \{\boldsymbol{\xi}, \boldsymbol{\lambda}\}$, then the density of Q^{*n} with respect to ν is denoted f^{*n}.

Discrete Distributions

A distribution $Q : \mathcal{B}(\mathbf{R}) \to [0, 1]$ is *discrete* if there exists a countable set $S \in \mathcal{B}(\mathbf{R})$ satisfying $Q[S] = 1$. If $Q[\mathbf{N}_0] = 1$, then Q is absolutely continuous with respect to $\boldsymbol{\xi}$. For detailed information on discrete distributions, see Johnson and Kotz [1969] and Johnson, Kotz, and Kemp [1992].

The Binomial Distribution

For $m \in \mathbf{N}$ and $\vartheta \in (0,1)$, the *binomial distribution* $\mathbf{B}(m,\vartheta)$ is defined to be the distribution Q satisfying

$$Q[\{x\}] \;=\; \binom{m}{x} \vartheta^x (1-\vartheta)^{m-x}$$

for all $x \in \{0,1,\ldots,m\}$.

Expectation:

$$E[Q] \;=\; m\vartheta$$

Variance:

$$\text{var}\,[Q] \;=\; m\vartheta(1-\vartheta)$$

Characteristic function:

$$\varphi_Q(z) \;=\; ((1-\vartheta) + \vartheta e^{iz})^m$$

Moment generating function:

$$M_Q(z) \;=\; ((1-\vartheta) + \vartheta e^z)^m$$

Probability generating function:

$$m_Q(z) \;=\; ((1-\vartheta) + \vartheta z)^m$$

Special case: The *Bernoulli distribution* $\mathbf{B}(\vartheta) := \mathbf{B}(1,\vartheta)$.

The Delaporte Distribution

For $\alpha, \beta \in (0,\infty)$ and $\vartheta \in (0,1)$, the *Delaporte distribution* $\mathbf{Del}(\alpha,\beta,\vartheta)$ is defined to be the distribution

$$Q \;:=\; \mathbf{P}(\alpha) * \mathbf{NB}(\beta,\vartheta) \,.$$

The Geometric Distribution

For $m \in \mathbf{N}$ and $\vartheta \in (0,1)$, the *geometric distribution* $\mathbf{Geo}(m,\vartheta)$ is defined to be the distribution

$$Q \;:=\; \delta_m * \mathbf{NB}(m,\vartheta) \,.$$

Special case: The one–parameter geometric distribution $\mathbf{Geo}(\vartheta) := \mathbf{Geo}(1,\vartheta)$.

The Logarithmic Distribution

For $\vartheta \in (0,1)$, the *logarithmic distribution* $\mathbf{Log}(\vartheta)$ is defined to be the distribution Q satisfying

$$Q[\{x\}] \;=\; \frac{1}{|\log(1-\vartheta)|} \frac{\vartheta^x}{x}$$

for all $x \in \mathbf{N}$.

Expectation:

$$E[Q] \;=\; \frac{1}{|\log(1-\vartheta)|} \frac{\vartheta}{1-\vartheta}$$

Variance:

$$\mathrm{var}\,[Q] \;=\; \frac{|\log(1-\vartheta)| - \vartheta}{|\log(1-\vartheta)|^2} \frac{\vartheta}{(1-\vartheta)^2}$$

Characteristic function:

$$\varphi_Q(z) \;=\; \frac{\log(1-\vartheta e^{iz})}{\log(1-\vartheta)}$$

Moment generating function for $z \in (-\infty, -\log(\vartheta))$:

$$M_Q(z) \;=\; \frac{\log(1-\vartheta e^z)}{\log(1-\vartheta)}$$

Probability generating function:

$$m_Q(z) \;=\; \frac{\log(1-\vartheta z)}{\log(1-\vartheta)}$$

The Negativebinomial Distribution

For $\alpha \in (0,\infty)$ and $\vartheta \in (0,1)$, the *negativebinomial distribution* $\mathbf{NB}(\alpha,\vartheta)$ is defined to be the distribution Q satisfying

$$Q[\{x\}] \;=\; \binom{\alpha + x - 1}{x} \vartheta^\alpha (1-\vartheta)^x$$

for all $x \in \mathbf{N}_0$.

Expectation:

$$E[Q] \;=\; \alpha \frac{1-\vartheta}{\vartheta}$$

Variance:

$$\text{var}\,[Q] \;\; = \;\; \alpha\,\frac{1-\vartheta}{\vartheta^2}$$

Characteristic function:

$$\varphi_Q(z) \;\; = \;\; \left(\frac{\vartheta}{1-(1-\vartheta)e^{iz}}\right)^{\alpha}$$

Moment generating function for $z \in (-\infty, -\log(1-\vartheta))$:

$$M_Q(z) \;\; = \;\; \left(\frac{\vartheta}{1-(1-\vartheta)e^z}\right)^{\alpha}$$

Probability generating function:

$$m_Q(z) \;\; = \;\; \left(\frac{\vartheta}{1-(1-\vartheta)z}\right)^{\alpha}$$

Special case: The *Pascal distribution* $\mathbf{NB}(m,\vartheta)$ with $m \in \mathbf{N}$.

The Negativehypergeometric (or Pólya–Eggenberger) Distribution

For $m \in \mathbf{N}$ and $\alpha, \beta \in (0,\infty)$, the *negativehypergeometric distribution* or *Pólya-Eggenberger distribution* $\mathbf{NH}(m,\alpha,\beta)$ is defined to be the distribution Q satisfying

$$Q[\{x\}] \;\; = \;\; \binom{\alpha+x-1}{x}\binom{\beta+m-x-1}{m-x}\binom{\alpha+\beta+m-1}{m}^{-1}$$

for all $x \in \{0,\ldots,m\}$.

Expectation:

$$E[Q] \;\; = \;\; m\,\frac{\alpha}{\alpha+\beta}$$

Variance:

$$\text{var}\,[Q] \;\; = \;\; m\,\frac{\alpha\beta}{(\alpha+\beta)^2}\,\frac{\alpha+\beta+m}{\alpha+\beta+1}$$

The Poisson Distribution

For $\alpha \in (0,\infty)$, the *Poisson distribution* $\mathbf{P}(\alpha)$ is defined to be the distribution Q satisfying

$$Q[\{x\}] \;\; = \;\; e^{-\alpha}\,\frac{\alpha^x}{x!}$$

for all $x \in \mathbf{N}_0$.

Expectation:

$$E[Q] \;=\; \alpha$$

Variance:

$$\text{var}\,[Q] \;=\; \alpha$$

Characteristic function:

$$\varphi_Q(z) \;=\; e^{\alpha(e^{iz}-1)}$$

Moment generating function:

$$M_Q(z) \;=\; e^{\alpha(e^{z}-1)}$$

Probability generating function:

$$m_Q(z) \;=\; e^{\alpha(z-1)}$$

Continuous Distributions

A distribution $Q : \mathcal{B}(\mathbf{R}) \to [0,1]$ is *continuous* if it is absolutely continuous with respect to $\boldsymbol{\lambda}$. For detailed information on continuous distributions, see Johnson and Kotz [1970a, 1970b].

The Beta Distribution

For $\alpha, \beta \in (0, \infty)$, the *beta distribution* $\mathbf{Be}(\alpha, \beta)$ is defined to be the distribution

$$Q \;:=\; \int \frac{1}{\mathrm{B}(\alpha, \beta)} \, x^{\alpha-1} \, (1-x)^{\beta-1} \, \chi_{(0,1)}(x) \, d\boldsymbol{\lambda}(x) \,.$$

Expectation:

$$E[Q] \;=\; \frac{\alpha}{\alpha + \beta}$$

Variance:

$$\text{var}\,[Q] \;=\; \frac{\alpha\beta}{(\alpha + \beta)^2(\alpha + \beta + 1)}$$

Special case: The *uniform distribution* $\mathbf{U}(0,1) := \mathbf{Be}(1,1)$.

The Gamma Distribution (Two Parameters)

For $\alpha, \beta \in (0, \infty)$, the *gamma distribution* $\mathbf{Ga}(\alpha, \beta)$ is defined to be the distribution

$$Q := \int \frac{\alpha^\beta}{\Gamma(\beta)} e^{-\alpha x} x^{\beta-1} \chi_{(0,\infty)}(x) \, d\boldsymbol{\lambda}(x) .$$

Expectation:

$$E[Q] = \frac{\beta}{\alpha}$$

Variance:

$$\mathrm{var}\,[Q] = \frac{\beta}{\alpha^2}$$

Characteristic function:

$$\varphi_Q(z) = \left(\frac{\alpha}{\alpha - iz}\right)^\beta$$

Moment generating function for $z \in (-\infty, \alpha)$:

$$M_Q(z) = \left(\frac{\alpha}{\alpha - z}\right)^\beta$$

Special cases:
- The *Erlang distribution* $\mathbf{Ga}(\alpha, m)$ with $m \in \mathbf{N}$.
- The *exponential distribution* $\mathbf{Exp}(\alpha) := \mathbf{Ga}(\alpha, 1)$.
- The *chi-square distribution* $\boldsymbol{\chi}_m^2 := \mathbf{Ga}(1/2, m/2)$ with $m \in \mathbf{N}$.

The Gamma Distribution (Three Parameters)

For $\alpha, \beta \in (0, \infty)$ and $\gamma \in \mathbf{R}$, the *gamma distribution* $\mathbf{Ga}(\alpha, \beta, \gamma)$ is defined to be the distribution

$$Q := \boldsymbol{\delta}_\gamma * \mathbf{Ga}(\alpha, \beta) .$$

Special case: The two–parameter gamma distribution $\mathbf{Ga}(\alpha, \beta) = \mathbf{Ga}(\alpha, \beta, 0)$.

The Pareto Distribution

For $\alpha, \beta \in (0, \infty)$, the *Pareto distribution* $\mathbf{Par}(\alpha, \beta)$ is defined to be the distribution

$$Q := \int \frac{\beta}{\alpha} \left(\frac{\alpha}{\alpha + x}\right)^{\beta+1} \chi_{(0,\infty)}(x) \, d\boldsymbol{\lambda}(x) .$$

Bibliography

Adelson, R. M.

[1966] *Compound Poisson distributions.* Oper. Res. Quart. **17**, 73–75.

Albrecht, P.

[1981] *Dynamische statistische Entscheidungsverfahren für Schadenzahlprozesse.* Karlsruhe: Verlag Versicherungswirtschaft.

[1985] *Mixed Poisson process.* In: *Encyclopedia of Statistical Sciences,* Vol. *6,* pp. 556–559. New York – Chichester: Wiley.

Aliprantis, C. D., and Burkinshaw, O.

[1990] *Principles of Real Analysis.* Second Edition. Boston – New York: Academic Press.

Alsmeyer, G.

[1991] *Erneuerungstheorie.* Stuttgart: Teubner.

Ambagaspitiya, R. S.

[1995] *A family of discrete distributions.* Insurance Math. Econom. **16**, 107–127.

Ambagaspitiya, R. S., and Balakrishnan, N.

[1994] *On the compound generalized Poisson distributions.* ASTIN Bull. **24**, 255–263.

Ammeter, H.

[1948] *A generalization of the collective theory of risk in regard to fluctuating basic-probabilities.* Scand. Actuar. J. **31**, 171–198.

Azlarov, T. A., and Volodin, N. A.

[1986] *Characterization Problems Associated with the Exponential Distribution.* Berlin – Heidelberg – New York: Springer.

Balakrishnan, N. (see R. S. Ambagaspitiya)

Bauer, H.

[1991] *Wahrscheinlichkeitstheorie.* 4. Auflage. Berlin: DeGruyter.

[1992] *Maß- und Integrationstheorie.* 2. Auflage. Berlin: DeGruyter.

Bauwelinckx, T. (see M. J. Goovaerts)

Beirlant, J., and Teugels, J. L.

[1992]   *Modeling large claims in non-life insurance.* Insurance Math. Econom. **11** 17–29.

Bichsel, F.

[1964]   *Erfahrungs–Tarifierung in der Motorfahrzeughaftpflicht–Versicherung.* Mitt. SVVM **64**, 119–130.

Billingsley, P.

[1995]   *Probability and Measure.* Third Edition. New York – Chichester: Wiley.

Björk, T., and Grandell, J.

[1988]   *Exponential inequalities for ruin probabilities.* Scand. Actuar. J., 77–111.

Bowers, N. L., Gerber, H. U., Hickman, J. C., Jones, D. A., and Nesbitt, C. J.

[1986]   *Actuarial Mathematics.* Itasca (Illinois): The Society of Actuaries.

Brémaud, P.

[1981]   *Point Processes and Queues.* Berlin – Heidelberg – New York: Springer.

Bühlmann, H.

[1970]   *Mathematical Methods in Risk Theory.* Berlin – Heidelberg – New York: Springer.

[1972]   *Ruinwahrscheinlichkeit bei erfahrungstarifiertem Portefeuille.* Mitt. SVVM **72**, 211–224.

Burkinshaw, O. (see C. D. Aliprantis)

Chow, Y. S., and Teicher, H.

[1988]   *Probability Theory – Independence, Interchangeability, Martingales.* Second Edition. Berlin – Heidelberg – New York: Springer.

Cox, D. R., and Isham, V.

[1980]   *Point Processes.* London – New York: Chapman and Hall.

Csörgö, M., and Steinebach, J.

[1991]   *On the estimation of the adjustment coefficient in risk theory via intermediate order statistics.* Insurance Math. Econom. **10**, 37–50.

Dassios, A., and Embrechts, P.

[1989]   *Martingales and insurance risk.* Comm. Statist. Stoch. Models **5**, 181–217.

Daykin, C. D., Pentikäinen, T., and Pesonen, M.

[1994]   *Practical Risk Theory for Actuaries.* London – New York: Chapman and Hall.

DeDominicis, R. (see J. Janssen)

Deheuvels, P., and Steinebach, J.

[1990] *On some alternative estimates of the adjustment coefficient.* Scand. Actuar. J., 135–159.

Delbaen, F., and Haezendonck, J.

[1985] *Inversed martingales in risk theory.* Insurance Math. Econom. **4**, 201–206.

Delaporte, P. J.

[1960] *Un problème de tarification de l'assurance accidents d'automobiles examiné par la statistique mathématique.* In: *Transactions of the International Congress of Actuaries, Vol. 2*, pp. 121–135.

[1965] *Tarification du risque individuel d'accidents d'automobiles par la prime modelée sur le risque.* ASTIN Bull. **3**, 251–271.

DePril, N.

[1986] *Moments of a class of compound distributions.* Scand. Actuar. J., 117–120.

Derron, M.

[1962] *Mathematische Probleme der Automobilversicherung.* Mitt. SVVM **62**, 103–123.

DeVylder, F. (see also M. J. Goovaerts)

[1977] *Martingales and ruin in a dynamical risk process.* Scand. Actuar. J., 217–225.

Dickson, D. C. M.

[1991] *The probability of ultimate ruin with a variable premium loading – a special case.* Scand. Actuar. J., 75–86.

Dienst, H. R. (ed.)

[1988] *Mathematische Verfahren der Rückversicherung.* Karlsruhe: Verlag Versicherungswirtschaft.

Dubourdieu, J.

[1938] *Remarques relatives à la théorie mathématique de l'assurance–accidents.* Bull. Inst. Actu. Franç. **44**, 79–126.

Embrechts, P. (see also A. Dassios)

Embrechts, P., and Mikosch, T.

[1991] *A bootstrap procedure for estimating the adjustment coefficient.* Insurance Math. Econom. **10**, 181–190.

Embrechts, P., Grandell, J., and Schmidli, H.

[1993] *Finite–time Lundberg inequalities on the Cox case.* Scand. Actuar. J. 17–41.

Embrechts, P., and Schmidli, H.

[1994]   *Ruin estimation for a general insurance risk model.* Adv. Appl. Probab. **26**, 404–422.

Embrechts, P., and Veraverbeke, N.

[1982]   *Estimates for the probability of ruin with special emphasis on the possibility of large claims.* Insurance Math. Econom. **1**, 55–72.

Embrechts, P., and Villaseñor, J. A.

[1988]   *Ruin estimates for large claims.* Insurance Math. Econom. **7**, 269–274.

Galambos, J., and Kotz, S.

[1978]   *Characterizations of Probability Distributions.* Berlin – Heidelberg – New York: Springer.

Gerathewohl, K.

[1976]   *Rückversicherung – Grundlagen und Praxis, Band 1.* Karlsruhe: Verlag Versicherungswirtschaft.

[1979]   *Rückversicherung – Grundlagen und Praxis, Band 2.* Karlsruhe: Verlag Versicherungswirtschaft.

Gerber, H. U. (see also N. L. Bowers)

[1973]   *Martingales in risk theory.* Mitt. SVVM **73**, 205–216.

[1979]   *An Introduction to Mathematical Risk Theory.* Homewood (Illinois): Irwin.

[1983]   *On the asymptotic behaviour of the mixed Poisson process.* Scand. Actuar. J. 256.

[1986]   *Lebensversicherungsmathematik.* Berlin – Heidelberg – New York: Springer.

[1990]   *Life Insurance Mathematics.* Berlin – Heidelberg – New York: Springer.

[1991]   *From the generalized gamma to the generalized negative binomial distribution.* Insurance Math. Econom. **10**, 303–309.

[1994]   *Martingales and tail probabilities.* ASTIN Bull. **24**, 145–146.

[1995]   *Life Insurance Mathematics.* Second Edition. Berlin – Heidelberg – New York: Springer.

Goovaerts, M. J. (see also R. Kaas, B. Kling, J. T. Runnenburg)

Goovaerts, M. J., DeVylder, F., and Haezendonck, J.

[1984]   *Insurance Premiums.* Amsterdam – New York: North–Holland.

Goovaerts, M. J., and Kaas, R.

[1991]   *Evaluating compound generalized Poisson distributions recursively.* ASTIN Bull. **21**, 193–198.

Goovaerts, M. J., Kaas, R., Van Heerwaarden, A. E., and Bauwelinckx, T.

[1990]   *Effective Actuarial Methods.* Amsterdam – New York: North–Holland.

Grandell, J. (see also T. Björk, P. Embrechts)

[1977]   *Point processes and random measures.* Adv. Appl. Probab. **9**, 502–526, 861.

[1991]   *Aspects of Risk Theory.* Berlin – Heidelberg – New York: Springer.

[1995]   *Mixed Poisson Processes.* Manuscript.

Gurland, J. (see R. Shumway)

Gut, A.

[1988]   *Stopped Random Walks – Limit Theorems and Applications.* Berlin – Heidelberg – New York: Springer.

Gut, A., and Schmidt, K. D.

[1983]   *Amarts and Set Function Processes.* Berlin – Heidelberg – New York: Springer.

Haezendonck, J. (see F. Delbaen, M. J. Goovaerts)

Heilmann, W. R.

[1987]   *Grundbegriffe der Risikotheorie.* Karlsruhe: Verlag Versicherungswirtschaft.

[1988]   *Fundamentals of Risk Theory.* Karlsruhe: Verlag Versicherungswirtschaft.

Helbig, M., and Milbrodt, H.

[1995]   *Mathematik der Personenversicherung.* Manuscript.

Heller, U. (see D. Pfeifer)

Helten, E., and Sterk, H. P.

[1976]   *Zur Typisierung von Schadensummenverteilungen.* Z. Versicherungswissenschaft **64**, 113–120.

Herkenrath, U.

[1986]   *On the estimation of the adjustment coefficient in risk theory by means of stochastic approximation procedures.* Insurance Math. Econom. **5**, 305–313.

Hesselager, O.

[1994]   *A recursive procedure for calculation of some compound distributions.* ASTIN Bull. **24**, 19–32.

Hickman, J. C. (see N. L. Bowers)

Hipp, C., and Michel, R.

[1990]   *Risikotheorie: Stochastische Modelle und Statistische Methoden.* Karlsruhe: Verlag Versicherungswirtschaft.

Hofmann, M.

[1955]   *Über zusammengesetzte Poisson–Prozesse und ihre Anwendungen in der Unfallversicherung.* Mitt. SVVM **55**, 499–575.

Isham, V. (see D. R. Cox)

Janssen, J.

[1977]   *The semi–Markov model in risk theory.* In: *Advances in Operations Research*, pp. 613–621. Amsterdam – New York: North–Holland.

[1982]   *Stationary semi–Markov models in risk and queuing theories.* Scand. Actuar. J., 199–210.

[1984]   *Semi–Markov models in economics and insurance.* In: *Premium Calculation in Insurance*, pp. 241–261. Dordrecht – Boston: Reidel.

Janssen, J., and DeDominicis, R.

[1984]   *Finite non–homogeneous semi–Markov processes: Theoretical and computational aspects.* Insurance Math. Econom. **3**, 157–165; **4**, 295 (1985).

Jewell, W. S. (see B. Sundt)

Johnson, N. L., and Kotz, S.

[1969]   *Distributions in Statistics: Discrete Distributions.* New York – Chichester: Wiley.

[1970a] *Distributions in Statistics: Continuous Univariate Distributions, Vol. 1.* New York – Chichester: Wiley.

[1970b] *Distributions in Statistics: Continuous Univariate Distributions, Vol. 2.* New York – Chichester: Wiley.

Johnson, N. L., Kotz, S., and Kemp, A. W.

[1992]   *Univariate Discrete Distributions.* Second Edition. New York – Chichester: Wiley.

Jones, D. A. (see N. L. Bowers)

Kaas, R. (see also M. J. Goovaerts)

Kaas, R., and Goovaerts, M. J.

[1985]   *Computing moments of compound distributions.* Scand. Actuar. J., 35 -38.

[1986]   *Bounds on stop–loss premiums for compound distributions.* ASTIN Bull. **16**, 13–18.

Kallenberg, O.

[1983]   *Random Measures.* London – Oxford: Academic Press.

Karr, A. F.

[1991]   *Point Processes and Their Statistical Inference.* Second Edition. New York – Basel: Dekker.

Kemp, A. W. (see N. L. Johnson)

Kerstan, J. (see also K. Matthes)

Kerstan, J., Matthes, K., and Mecke, J.

[1974]   *Unbegrenzt teilbare Punktprozesse.* Berlin: Akademie–Verlag.

Kingman, J. F. C.

[1993]   *Poisson Processes.* Oxford: Clarendon Press.

Kling, B., and Goovaerts, M. J.

[1993]   *A note on compound generalized distributions.* Scand. Actuar. J., 60–72.

Klüppelberg, C.

[1989]   *Estimation of ruin probabilities by means of hazard rates.* Insurance Math. Econom. **8**, 279–285.

König, D., and Schmidt, V.

[1992]   *Zufällige Punktprozesse.* Stuttgart: Teubner.

Kotz, S. (see J. Galambos, N. L. Johnson)

Kupper, J.

[1962]   *Wahrscheinlichkeitstheoretische Modelle in der Schadenversicherung. Teil I: Die Schadenzahl.* Blätter DGVM **5**, 451–503.

Lemaire, J.

[1985]   *Automobile Insurance.* Boston – Dordrecht – Lancaster: Kluwer–Nijhoff.

Letta, G.

[1984]   *Sur une caractérisation classique du processus de Poisson.* Expos. Math. **2**, 179–182.

Lin, X. (see G. E. Willmot)

Lundberg, O.

[1940]   *On Random Processes and Their Application to Sickness and Accident Statistics.* Uppsala: Almqvist and Wiksells.

Mammitzsch, V.

[1983]   *Ein einfacher Beweis zur Konstruktion der operationalen Zeit.* Blätter DGVM **16**, 1–3.

[1984]   *Operational time: A short and simple existence proof.* In: *Premium Calculation in Insurance*, pp. 461-465. Dordrecht -- Boston: Reidel.

[1986]   *A note on the adjustment coefficient in ruin theory.* Insurance Math. Econom. **5**, 147–149.

Mathar, R., and Pfeifer, D.

[1990]   *Stochastik für Informatiker.* Stuttgart: Teubner.

Matthes, K. (see also J. Kerstan)

Matthes, K., Kerstan, J., and Mecke, J.

[1978]   *Infinitely Divisible Point Processes.* New York – Chichester: Wiley.

Mecke, J. (see J. Kerstan, K. Matthes)

Michel, R. (see also C. Hipp)

[1993a]   *On probabilities of large claims that are compound Poisson distributed.* Blätter DGVM **21**, 207–211.

[1993b]   *Ein individuell-kollektives Modell für Schadenzahl-Verteilungen.*   Mitt. SVVM, 75–93.

Mikosch, T. (see P. Embrechts)

Milbrodt, H. (see M. Helbig)

Møller, C. M.

[1992]   *Martingale results in risk theory with a view to ruin probabilities and diffusions.* Scand. Actuar. J. 123–139.

[1995]   *Stochastic differential equations for ruin probabilities.* J. Appl. Probab. **32**, 74–89.

Nesbitt, C. J. (see N. L. Bowers)

Neveu, J.

[1972]   *Martingales à Temps Discret.* Paris: Masson.

[1977]   *Processus ponctuels.* In: *Ecole d'Eté de Probabilités de Saint-Flour VI*, pp. 249–445. Berlin – Heidelberg – New York: Springer.

Nollau, V.

[1978]   *Semi-Markovsche Prozesse.* Berlin: Akademie-Verlag.

Norberg, R.

[1990]   *Risk theory and its statistics environment.* Statistics **21**, 273–299.

Panjer, H. H. (see also S. Wang, G. E. Willmot)

[1981]   *Recursive evaluation of a family of compound distributions.* ASTIN Bull. **12**, 22–26.

Panjer, H. H., and Wang, S.

[1993]   *On the stability of recursive formulas.* ASTIN Bull. **23**, 227–258.

Panjer, H. H., and Willmot, G. E.

[1981]   *Finite sum evaluation of the negativebinomial–exponential model.* ASTIN Bull. **12**, 133–137.

[1982]   *Recursions for compound distributions.* ASTIN Bull. **13**, 1–11.

[1986]   *Computational aspects of recursive evaluation of compound distributions.* Insurance Math. Econom. **5**, 113–116.

[1992]   *Insurance Risk Models.* Schaumburg (Illinois): Society of Actuaries.

Pentikäinen, T. (see C. D. Daykin)

Pesonen, M. (see C. D. Daykin)

Pfeifer, D. (see also R. Mathar)

[1982a] *The structure of elementary birth processes.* J. Appl. Probab. **19**, 664-667.

[1982b] *An alternative proof of a limit theorem for the Pólya-Lundberg process.* Scand. Actuar. J. 176-178.

[1986] *Pólya-Lundberg process.* In: *Encyclopedia of Statistical Sciences, Vol. 7,* pp. 63-65. New York - Chichester: Wiley.

[1987] *Martingale characteristics of mixed Poisson processes.* Blätter DGVM **18**, 107-100.

Pfeifer, D., and Heller, U.

[1987] *A martingale characterization of mixed Poisson processes.* J. Appl. Probab. **24**, 246-251.

Quenouille, M. H.

[1949] *A relation between the logarithmic, Poisson and negative binomial series.* Biometrics **5**, 162-164.

Reiss, R. D.

[1993] *A Course on Point Processes.* Berlin - Heidelberg - New York: Springer.

Resnick, S. I.

[1992] *Adventures in Stochastic Processes.* Boston - Basel - Berlin: Birkhäuser.

Rhiel, R.

[1985] *Zur Berechnung von Erwartungswerten und Varianzen von zufälligen Summen in der kollektiven Risikotheorie.* Blätter DGVM **17**, 15-18.

[1986] *A general model in risk theory. An application of modern martingale theory. Part one: Theoretic foundations.* Blätter DGVM **17**, 401-428.

[1987] *A general model in risk theory. An application of modern martingale theory. Part two: Applications.* Blätter DGVM **18**, 1-19.

Runnenburg, J. T., and Goovaerts, M. J.

[1985] *Bounds on compound distributions and stop-loss premiums.* Insurance Math. Econom. **4**, 287-293.

Ruohonen, M.

[1988] *On a model for the claim number process.* ASTIN Bull. **18**, 57-68.

Scheike, T. H.

[1992] *A general risk process and its properties.* J. Appl. Probab. **29**, 73-81.

Schmidli, H. (see also P. Embrechts)

[1995] *Cramér-Lundberg approximations for ruin probabilities of risk processes perturbed by diffusion.* Insurance Math. Econom. **16**, 135-149.

Schmidt, K. D. (see also A. Gut)

[1989] *A note on positive supermartingales in ruin theory.* Blätter DGVM **19**, 129–132.

[1992] *Stochastische Modellierung in der Erfahrungstarifierung.* Blätter DGVM **20**, 441–455.

Schmidt, V. (see D. König)

Schröter, K. J.

[1990] *On a family of counting distributions and recursions for related compound distributions.* Scand. Actuar. J., 161–175.

[1995] *Verfahren zur Approximation der Gesamtschadenverteilung – Systematisierung, Techniken und Vergleiche.* Karlsruhe: Verlag Versicherungswirtschaft.

Seal, H. L.

[1980] *Survival probabilities based on Pareto claim distributions.* ASTIN Bull. **11**, 61–71.

[1983] *The Poisson process: Its failure in risk theory.* Insurance Math. Econom. **2**, 287–288.

Shumway, R., and Gurland, J.

[1960] *A fitting procedure for some generalized Poisson distributions.* Scand. Actuar. J. **43**, 87–108.

Sobrero, M. (see S. Wang)

Steinebach, J. (see also M. Csörgö, P. Deheuvels)

[1993] *Zur Schätzung der Cramér-Lundberg-Schranke, wenn kein Anpassungskoeffizient existiert.* In: *Geld, Finanzwirtschaft, Banken und Versicherungen*, pp. 715–723. Karlsruhe: Verlag Versicherungswirtschaft

Sterk, H. P. (see also E. Helten)

[1979] *Selbstbeteiligung unter risikotheoretischen Aspekten.* Karlsruhe: Verlag Versicherungswirtschaft.

[1980] *Risikotheoretische Aspekte von Selbstbeteiligungen.* Blätter DGVM **14**, 413–426.

[1988] *Selbstbeteiligung.* In: *Handwörterbuch der Versicherung*, pp. 775–780. Karlsruhe: Verlag Versicherungswirtschaft.

Störmer, H.

[1969] *Zur Überlagerung von Erneuerungsprozessen.* Z. Wahrscheinlichkeitstheorie verw. Gebiete **13**, 9–24.

[1970] *Semi-Markov-Prozesse mit endlich vielen Zuständen.* Berlin – Heidelberg – New York: Springer.

Straub, E.

[1988]   *Non–Life Insurance Mathematics.*   Berlin – Heidelberg – New York: Springer.

Sundt, B. (see also G. E. Willmot)

[1984]   *An Introduction to Non–Life Insurance Mathematics.* First Edition. Karlsruhe: Verlag Versicherungswirtschaft.

[1991]   *An Introduction to Non–Life Insurance Mathematics.* Second Edition. Karlsruhe: Verlag Versicherungswirtschaft.

[1992]   *On some extensions of Panjer's class of counting distributions.* ASTIN Bull. **22**, 61–80.

[1993]   *An Introduction to Non–Life Insurance Mathematics.* Third Edition. Karlsruhe: Verlag Versicherungswirtschaft.

Sundt, B., and Jewell, W. S.

[1981]   *Further results on recursive evaluation of compound distributions.* ASTIN Bull. **12**, 27–39.

Teicher, H. (see Y. S. Chow)

Teugels, J. L. (see J. Beirlant)

Thorin, O., and Wikstad, N.

[1977]   *Calculation of ruin probabilities when the claim distribution is lognormal.* ASTIN Bull. **9**, 231–246.

Thyrion, P.

[1960]   *Contribution à l'étude du bonus pour non sinistre en assurance automobile.* ASTIN Bull. **1**, 142–162.

Tröblinger, A.

[1961]   *Mathematische Untersuchungen zur Beitragsrückgewähr in der Kraftfahrversicherung.* Blätter DGVM **5**, 327–348.

[1975]   *Analyse und Prognose des Schadenbedarfs in der Kraftfahrzeughaftpflichtversicherung.* Karlsruhe: Verlag Versicherungswirtschaft.

Van Heerwaarden, A. E. (see M. J. Goovaerts)

Veraverbeke, N. (see P. Embrechts)

Villaseñor, J. A. (see P. Embrechts)

Volodin, N. A. (see T. A. Azlarov)

Wang, S. (see also H. H. Panjer)

Wang, S., and Panjer, H. H.

[1994]   *Proportional convergence and tail–cutting techniques in evaluating aggregate claim distributions.* Insurance Math. Econom. **14**, 129–138.

Wang, S., and Sobrero, M.

[1994] *Further results on Hesselager's recursive procedure for calculation of some compound distributions.* ASTIN Bull. **24**, 161-166.

Watanabe, S.

[1964] *On discontinuous additive functionals and Lévy measures of a Markov process.* Japan. J. Math. **34**, 53-70.

Wikstad, N. (see O. Thorin)

Willmot, G. E. (see also H. H. Panjer)

[1986] *Mixed compound Poisson distributions.* ASTIN Bull. **16**, S59-S80.

[1988] *Sundt and Jewell's family of discrete distributions.* ASTIN Bull. **18**, 17-30.

Willmot, G. E., and Lin, X.

[1994] *Lundberg bounds on the tails of compound distributions.* J. Appl. Probab. **31**, 743-756.

Willmot, G. E., and Panjer, H. H.

[1987] *Difference equation approaches in evaluation of compound distributions.* Insurance Math. Econom. **6**, 43-56.

Willmot, G. E., and Sundt, B.

[1989] *On evaluation of the Delaporte distribution and related distributions.* Scand. Actuar. J., 101-113.

Wolfsdorf, K.

[1986] *Versicherungsmathematik. Teil 1: Personenversicherung.* Stuttgart: Teubner.

[1988] *Versicherungsmathematik. Teil 2: Theoretische Grundlagen.* Stuttgart: Teubner.

Wolthuis, H.

[1994] *Life Insurance Mathematics – The Markovian Approach.* Brussels: CAIRE.

List of Symbols

Numbers and Vectors

$\mathbf{N}$	the set $\{1, 2, \ldots\}$
$\mathbf{N}_0$	the set $\{0, 1, 2, \ldots\}$
$\mathbf{Q}$	the set of rational numbers
$\mathbf{R}$	the set of real numbers
$\mathbf{R}_+$	the interval $[0, \infty)$
$\mathbf{R}^n$	the n–dimensional Euclidean space

Sets

χ_A	indicator function of the set A
$\sum_{i \in \mathbf{I}} A_i$	union of the (pairwise) disjoint family $\{A_i\}_{i \in \mathbf{I}}$

σ–Algebras

$\sigma(\mathcal{E})$	σ–algebra generated by the class $\mathcal{E}$
$\sigma(\{Z_i\}_{i \in \mathbf{I}})$	σ–algebra generated by the family $\{Z_i\}_{i \in \mathbf{I}}$
$\mathcal{B}((0, \infty))$	Borel σ–algebra on $(0, \infty)$
$\mathcal{B}(\mathbf{R})$	Borel σ–algebra on $\mathbf{R}$
$\mathcal{B}(\mathbf{R}^n)$	Borel σ–algebra on $\mathbf{R}^n$

Measures

μ_Z	transformation of μ under Z
$\mu\|_{\mathcal{E}}$	restriction of μ to $\mathcal{E}$
$\mu \otimes \nu$	product measure of μ and ν
$\mu * \nu$	convolution of (the measures or densities) μ and ν
λ	Lebesgue measure on $\mathcal{B}(\mathbf{R})$
λ^n	Lebesgue measure on $\mathcal{B}(\mathbf{R}^n)$
ξ	counting measure concentrated on $\mathbf{N}_0$

Integrals

$\int_A f(x)\, d\mu(x)$	Lebesgue integral of f on A with respect to μ
$\int_a^c f(x)\, dx$	Riemann integral of f on $[a, c]$

Distributions

$\boldsymbol{\delta}_y$	Dirac distribution
$\mathbf{B}(\vartheta)$	Bernoulli distribution
$\mathbf{B}(m,\vartheta)$	binomial distribution
$\mathbf{Be}(\alpha,\beta)$	beta distribution
$\mathbf{C}(Q,R)$	compound distribution
$\mathbf{Del}(\alpha,\beta,\vartheta)$	Delaporte distribution
$\mathbf{Exp}(\alpha)$	exponential distribution
$\mathbf{Ga}(\alpha,\beta)$	gamma distribution
$\mathbf{Ga}(\alpha,\beta,\gamma)$	gamma distribution
$\mathbf{Geo}(\vartheta)$	geometric distribution
$\mathbf{Geo}(m,\vartheta)$	geometric distribution
$\mathbf{Log}(\vartheta)$	logarithmic distribution
$\mathbf{NB}(\alpha,\vartheta)$	negativebinomial distribution
$\mathbf{NH}(m,\alpha,\beta)$	negativehypergeometric distribution
$\mathbf{P}(\alpha)$	Poisson distribution
$\mathbf{Par}(\alpha,\beta)$	Pareto distribution

Probability

$E[Z]$	expectation of (the random variable or distribution) Z
$\text{var}\,[Z]$	variance of Z
$v[Z]$	coefficient of variation of Z
φ_Z	characteristic function of Z
M_Z	moment generating function of Z
m_Z	probability generating function of Z

Conditional Probability

$P_{Z|\Theta}$	conditional distribution of Z with respect to $\sigma(\Theta)$
$E(Z|\Theta)$	conditional expectation of Z with respect to $\sigma(\Theta)$
$\text{var}\,(Z|\Theta)$	conditional variance of Z with respect to $\sigma(\Theta)$

Stochastic processes

$\{G_n\}_{n\in\mathbf{N}}$	excess premium process
$\{N_t\}_{t\in\mathbf{R}_+}$	claim number process
$\{R_t\}_{t\in\mathbf{R}_+}$	reserve process
$\{S_t\}_{t\in\mathbf{R}_+}$	aggregate claims process
$\{T_n\}_{n\in\mathbf{N}_0}$	claim arrival process
$\{U_n\}_{n\in\mathbf{N}_0}$	modified reserve process
$\{W_n\}_{n\in\mathbf{N}}$	claim interarrival process
$\{X_n\}_{n\in\mathbf{N}}$	claim size process

Author Index

A

Adelson, R. M., 125
Albrecht, P., 101, 102
Aliprantis, C. D., 3
Alsmeyer, G., 42
Ambagaspitiya, R. S., 126
Ammeter, H., 125
Azlarov, T. A., 16

B

Balakrishnan, N., 126
Bauer, H., 3, 85
Bauwelinckx, T., 4
Beirlant, J., 170
Bichsel, F., 102
Billingsley, P., 3, 85
Björk, T., 169
Bowers, N. L., 4
Brémaud, P., 42
Bühlmann, H., 4, 84, 170
Burkinshaw, O., 3

C

Chow, Y. S., 85
Cox, D. R., 42
Csörgö, M., 170

D

Dassios, A., 169, 170
Daykin, C. D., 4
DeDominicis, R., 84

Deheuvels, P., 170
Delaporte, P. J., 102
Delbaen, F., 169
DePril, N., 126
Derron, M., 102
DeVylder, F., 126, 169, 170
Dickson, D. C. M., 170
Dienst, H. R., 154
Dubourdieu, J., 101, 102

E

Embrechts, P., 170

G

Galambos, J., 16
Gerathewohl, K., 154
Gerber, H. U., 4, 101, 102, 125, 126, 169
Goovaerts, M. J., 4, 124, 126
Grandell, J., 41, 42, 101, 169, 170
Gurland, J., 125
Gut, A., 42, 170

H

Haezendonck, J., 126, 169
Heilmann, W. R., 4
Helbig, M., 4
Heller, U., 101
Helten, E., 1
Herkenrath, U., 170
Hesselager, O., 125
Hickman, J. C., 4

Hipp, C., 4, 126
Hofmann, M., 101

I

Isham, V., 42

J

Janssen, J., 84
Jewell, W. S., 125
Johnson, N. L., 125, 175, 179
Jones, D. A., 4

K

Kaas, R., 4, 124, 126
Kallenberg, O., 42
Karr, A. F., 42
Kemp, A. W., 175
Kerstan, J., 41, 42
Kingman, J. F. C., 42
Kling, B., 126
Klüppelberg, C., 170
König, D., 42
Kotz, S., 16, 125, 175, 179
Kupper, J., 1, 102

L

Lemaire, J., 102
Letta, G., 42
Lin, X., 125
Lundberg, O., 101, 124

M

Mammitzsch, V., 84, 170
Mathar, R., 42
Matthes, K., 41, 42
Mecke, J., 41, 42
Michel, R., 4, 125, 126
Mikosch, T., 170
Milbrodt, H., 4
Møller, C. M., 170

N

Nesbitt, C. J., 4
Neveu, J., 42, 170
Nollau, V., 84
Norberg, R., 4

P

Panjer, H. H., 4, 124, 125, 126, 171
Pentikäinen, T., 4
Pesonen, M., 4
Pfeifer, D., 42, 101

Q

Quenouille, M. H., 124

R

Reiss, R. D., 42
Resnick, S. I., 42
Rhiel, R., 124, 169
Runnenburg, J. T., 124
Ruohonen, M., 102

S

Scheike, T. H., 126
Schmidli, H., 170
Schmidt, K. D., 101, 170
Schmidt, V., 42
Schröter, K. J., 125, 126
Seal, H. L., 101, 170
Shumway, R., 125
Sobrero, M., 126
Steinebach, J., 170
Sterk, H. P., 1, 154
Störmer, H., 84, 154
Straub, E., 4
Sundt, B., 4, 84, 101, 125, 126

T

Teicher, H., 85
Teugels, J. L., 170
Thorin, O., 170
Thyrion, P., 102
Tröblinger, A., 1, 102

V

Van Heerwaarden, A. E., 4
Veraverbeke, N., 170
Villaseñor, J. A., 170
Volodin, N. A., 16

W

Wang, S., 125
Watanabe, S., 42
Wikstad, N., 170
Willmot, G. E., 4, 124, 125, 126, 171
Wolfsdorf, K., 4
Wolthuis, H., 4

Subject Index

A

abstract portfolio, 100
adjustment coefficient, 164
admissible pair, 45
aggregate claims amount, 103, 109
aggregate claims process, 103
Ammeter transform, 114

B

Bernoulli distribution, 176
Bernoulli process, 40
beta distribution, 179
beta function, 171
binomial criterion, 88
binomial distribution, 176
binomial process, 40
binomial risk process, 132
bounded stopping time, 161

C

canonical filtration, 25
Cantelli's inequality, 113
Chapman–Kolmogorov equations, 46
characteristic function, 174
claim amount, 103
claim arrival process, 6
claim event, 6
claim interarrival process, 6
claim measure, 8
claim number, 17, 109
claim number process, 17
claim severity, 103

claim size process, 103
coefficient of variation, 173
compound binomial distribution, 111
compound distribution, 109
compound negativebinomial
        distribution, 111
compound Poisson distribution, 109
compound Poisson process, 107
conditionally independent
        increments, 86
conditionally stationary
        increments, 86
contagion, 78, 82, 83, 95
continuous distribution, 179
continuous time model, 9
convolution, 175
counting measure, 172
Cox process, 170

D

decomposition of a claim size
        process, 137
decomposition of a Poisson
        process, 133
decomposition of a Poisson risk
        process, 137
degenerate distribution, 172
Delaporte distribution, 176
DePril's recursion, 123
Dirac distribution, 172
discrete distribution, 175
discrete time model, 9
disjoint familiy, 4

distribution, 172
doubly stochastic Poisson process, 170

E

Erlang distribution, 180
essential infimum, 157
estimation, 98
exceptional null set, 6, 17
excess of loss reinsurance, 108
excess premium process, 158
expectation, 173
experience rating, 101
explosion, 7, 10, 75
exponential distribution, 180

F

failure rate, 50
filtration, 25
finite expectation, 173
finite moment, 173
finite time ruin probability, 170
Fourier transform, 174

G

gamma distribution, 180
gamma function, 171
generalized binomial coefficient, 172
geometric distribution, 176

H

heavy tailed distribution, 170
homogeneous claim number process, 47
homogeneous Poisson process, 23

I

increasing, 4
increment, 20, 105
independent increments, 23, 105
inhomogeneous Poisson process, 56
initial reserve, 155

insurer's portfolio, 100
intensity, 49, 56

K

Kolmogorov's inequality, 164

L

layer, 132
Lebesgue measure, 172
life insurance, 9
lifetime, 9
logarithmic distribution, 177
Lundberg's binomial criterion, 88
Lundberg's inequality, 165

M

Markov (claim number) process, 44
martingale, 25
maximal inequality, 161
memoryless distribution, 13
mixed binomial process, 92
mixed claim number process, 86
mixed Poisson process, 87
modified reserve process, 158
moment, 173
moment generating function, 174
multinomial criterion, 23, 60, 87
multiple life insurance, 9

N

negative contagion, 82, 83
negativebinomial distribution, 177
negativehypergeometric distribution, 178
net premium, 160
nondegenerate distribution, 172
number of claim events, 107
number of claims, 17
number of claims occurring at a claim
        event, 107
number of large claims, 108

number of occurred claims, 107, 108
number of reported claims, 107

O

occurrence time, 7
operational time, 76

P

Panjer's recursion, 122
Pareto distribution, 180
partition, 4
Pascal distribution, 178
point process, 41
Poisson distribution, 178
Poisson process, 23
Poisson risk process, 137
Pólya-Eggenberger distribution, 178
Pólya-Lundberg process, 93
positive, 4
positive adapted sequence, 164
positive contagion, 78, 82, 95
prediction, 38, 39, 97, 98
premium income, 155
premium intensity, 155
priority, 108
probability generating function, 174

R

regular claim number process, 48
reinsurance, 108
renewal process, 42
reserve process, 155
risk process, 127
ruin, 158
ruin problem, 156

S

safety loading, 160
semi-Markov process, 84
sequence of intensities, 49

single life insurance, 9
stationary increments, 23, 105
stopping time, 161
structure distribution, 86
structure parameter, 86
submartingale, 25
superadjustment coefficient, 164
supermartingale, 25
superposition of claim size
      processes, 148
superposition of Poisson processes, 141
superposition of Poisson risk
      processes, 149
survival function, 13

T

tail probability, 113
thinned binomial process, 108
thinned claim number process, 107, 128
thinned claim size process, 130
thinned Poisson process, 108
thinned risk process, 130, 132
time of death, 9
total number of claims, 107
transition probability, 46
transition rule, 45

V

variance, 173
variance decomposition, 86

W

waiting time, 7
Wald's identities, 111

X

XL reinsurance, 108

Z

zero-one law on explosion, 10, 75

Teubner Studienbücher
zur Statistik

Afflerbach: **Statistik Praktikum mit dem PC**
198 Seiten. DM/SFr 24,80 / ÖS 194,–

Behnen/Neuhaus: **Grundkurs Stochastik**
3. Aufl. 493 Seiten. DM/SFr 54,80 / ÖS 406,–

v. Collani: **Optimale Wareneingangskontrolle**
150 Seiten. DM/SFr 29,80 / ÖS 233,–

Dinges/Rost: **Prinzipien der Stochastik**
294 Seiten. DM/SFr 38,– / ÖS 296,–

Dufner/Jensen/Schumacher: **Statistik mit SAS**
398 Seiten. DM/SFr 42,– / ÖS 328,–

Floret: **Maß- und Integrationstheorie**
360 Seiten. DM/SFr 39,80 / ÖS 311,–

Kohlas: **Stochastische Methoden des Operations Research**
192 Seiten. DM/SFr 26,80 / ÖS 209,–

Lehn/Wegmann: **Einführung in die Statistik**
2. Aufl. 202 Seiten. DM/SFr 27,80 / ÖS 217,–

Lehn/Wegmann/Rettig: **Aufgabensammlung zur Einführung in die Statistik**
2. Aufl. 256 Seiten. DM/SFr 29,80 / ÖS 233,–

Schmitz: **Vorlesungen über Wahrscheinlichkeitstheorie**
VII, 424 Seiten. DM/SFr 56,80 / ÖS 421,–

Topsøe: **Informationstheorie**
88 Seiten. DM/SFr 19,80 / ÖS 155,–

Uhlmann: **Statistische Qualitätskontrolle**
2. Aufl. 292 Seiten. DM/SFr 39,– / ÖS 304,–

Witting: **Mathematische Statistik**
3. Aufl. 223 Seiten. DM/SFr 29,80 / ÖS 233,–

Wolfsdorf: **Versicherungsmathematik**
Teil 1: Personenversicherung
2. Aufl. ca. 350 Seiten. ca. DM/SFr 46,– / ÖS 341,–
Teil 2: Theoretische Grundlagen, Risikotheorie, Sachversicherung
399 Seiten. DM/SFr 39,80 / ÖS 311,–

Preisänderungen vorbehalten

B. G. Teubner Stuttgart · Leipzig

Teubner Skripten zur Mathematischen Stochastik

Herausgegeben von: **J. Lehn, N. Schmitz** und **W. Weil**

Alsmeyer: **Erneuerungstheorie**
331 Seiten. DM 44,– / ÖS 343,– / SFr 44,–

Behnen/Neuhaus: **Rank Tests with Estimated Scores and Their Application**
XII, 416 pages. DM 54,– / ÖS 421,– / SFr 54,–

von Collani: **The Economic Design of Control Charts**
XII, 171 pages. DM 29,– / ÖS 226,– / SFr 29,–

Irle: **Sequentialanalyse: Optimale sequentielle Tests**
VII, 176 Seiten. DM 29,– / ÖS 226,– / SFr 29,–

Kamps: **A Concept of Generalized Order Statistics**
210 pages. DM 39,80 / ÖS 311,– / SFr 39,80

König/Schmidt: **Zufällige Punktprozesse**
363 Seiten. DM 49,– / ÖS 382,– / SFr 49,–

Pfeifer: **Einführung in die Extremwertstatistik**
VIII, 199 Seiten. DM 34,– / ÖS 265,– / SFr 34,–

Pruscha: **Angewandte Methoden der Mathematischen Statistik**
2. Aufl. II, 412 Seiten. DM 62,– / ÖS 459,– / SFr 62,–

Rüschendorf: **Asymptotische Statistik**
IX, 225 Seiten. DM 34,– / ÖS 265,– / SFr 34,–

Schäl: **Markoffsche Entscheidungsprozesse**
XV, 182 Seiten. DM 29,– / ÖS 226,– / SFr 29,–

Schmidt: **Lectures on Risk Theory**
X, 200 Seiten. DM 44,80 / ÖS 332,– / SFr 44,80

Schneider/Weil: **Integralgeometrie**
VII, 222 Seiten. DM 34,- / ÖS 265,– / SFr 34,–

Preisänderungen vorbehalten.

B. G. Teubner Stuttgart · Leipzig